KV-412-101

Photoelectron Spectroscopy

AN INTRODUCTION TO ULTRAVIOLET PHOTOELECTRON
SPECTROSCOPY IN THE GAS PHASE

Photoelectron Spectroscopy

AN INTRODUCTION TO ULTRAVIOLET PHOTOELECTRON
SPECTROSCOPY IN THE GAS PHASE

J. H. D. ELAND, M.A., D.PHIL.
Research Lecturer of Christ Church, Oxford

LONDON
BUTTERWORTHS

THE BUTTERWORTH GROUP

ENGLAND
Butterworth & Co. (Publishers) Ltd.
London: 88 Kingsway WC2B 6AB

AUSTRALIA
Butterworths Pty Ltd.
Sydney: 586 Pacific Highway NSW 2067
Melbourne: 343 Little Collins Street 3000
Brisbane: 240 Queen Street 4000

CANADA
Butterworth & Co. (Canada) Ltd.
Toronto: 14 Curity Avenue 374

NEW ZEALAND
Butterworths of New Zealand Ltd.
Wellington: 26–28 Waring Taylor Street 1

SOUTH AFRICA
Butterworth & Co. (South Africa) (Pty) Ltd.
Durban: 152–154 Gale Street

First published 1974

© Butterworth & Co. (Publishers) Ltd., 1974

ISBN 0 408 70559 0

Filmset and printed in England by
Cox & Wyman Ltd., London, Fakenham and Reading

Preface

The photoelectron spectrometer will soon take its place in the laboratory beside the mass spectrometers, optical spectrometers and radio-frequency spectrometers that have become routine tools of the chemist and physicist. A new form of molecular spectroscopy naturally requires an incubation period in the hands of specialist physicists and physical chemists before it becomes useful in wider fields of chemistry, and photoelectron spectroscopy is now emerging from such a stage in its development. Sure signs of this emergence are the burgeoning of chemical applications of the technique and the availability of commercial photoelectron spectrometers with very high performance. At the same time, there is a lack of any textbook that covers the new technique at an advanced undergraduate or first year research level, and this I have attempted to provide. My aim has been to cover, at least qualitatively, almost all that a chemist needs to know in order to interpret a photoelectron spectrum with which he is confronted. The treatment is experimentally based and non-mathematical, but assumes some familiarity with other spectroscopic techniques and with the chemical applications of Group Theory.

The importance of photoelectron spectroscopy in the study of molecular electronic structure is now widely appreciated; its relevance to mass spectrometry and unimolecular reaction rate theory deserves more attention than it has hitherto received, and I hope that the inclusion of Chapter 7 on ionic dissociation will go some way to rectify this. Chapters 1 to 6 form a progressive introduction to photoelectron spectroscopy, and they are intended to be read sequentially, with a few possible exceptions. The more difficult topics in Sections 1.4.2, 3.4.3, 3.5 and 4.6 could be omitted on a first reading and Chapter 2, on experimental methods, may be referred to separately from the main text. The final chapter contains accounts of some selected applications of photoelectron spectroscopy in chemistry, and includes a sufficiently full reference list for these topics to be followed up in detail. Shorter reference lists are provided for all the other chapters and should serve as a key to the literature, but they are by no means a complete bibliography; often only the most recent papers on a particular subject are cited. In a rapidly advancing field such as this, it is impossible to write a completely up-to-date book, and the inclusion of new material had to stop at the end of 1972.

In preparing this book I have been helped by discussions with several scientists, and I should like to thank Dr. B. Brehm, Dr. M. S. Child, Dr. C. J. Danby, Professor E. Heilbronner and Mr. A. F. Orchard in particular. Dr. Brehm and Dr. Danby also read parts of the manuscript in draft and made suggestions for several necessary improvements. The typing was undertaken by Mrs. M. Long, and I am most grateful for her speed and cheerfulness in dealing with a difficult manuscript. Finally, I want to thank my wife, Ieva, for the immense amount of help she has given at every stage of the work.

JOHN H. D. ELAND

Contents

1 Principles of Photoelectron Spectroscopy

1.1 INTRODUCTION

When light of short wavelength interacts with free molecules, it can cause electrons to be ejected from the occupied molecular orbitals. Photoelectron spectroscopy is the study of these photoelectrons, whose energies, abundances and angular distributions are all characteristic of the individual molecular orbitals from which they originate. The experimental singling-out of individual molecular orbitals is the outstanding feature of photoelectron spectroscopy, and one which distinguishes it from all other methods of examining molecular electronic structure.

The quantity measured most directly in photoelectron spectroscopy is the ionization potential for the removal of electrons from different molecular orbitals. According to an approximation known as Koopmans' theorem, each ionization potential, I_j, is equal in magnitude to an orbital energy, ε_j:

$$I_j = -\varepsilon_j \qquad (1.1)$$

This is an approximation additional to those inherent in the molecular orbital model for many-electron systems, but it is a good and a very useful one. It means that the photoelectron spectrum of a molecule is a direct representation of the molecular orbital energy diagram, and the spectrum gives not only the orbital energies but also, less directly, the changes in molecular geometry caused by removal of one electron from each orbital. These changes reveal the character

1

of the orbitals, whether they are bonding, antibonding or non-bonding, and how their bonding power is localized in the molecules.

The photoelectron spectroscopy discussed in this book is based on photoionization brought about by radiation in the far ultraviolet region of the spectrum. It was discovered early in the 1960s independently by two groups, one led by Turner[1, 2] in London, the other by Vilesov[3] in Leningrad. Siegbahn[4] and his group at Uppsala had evolved a similar technique a little earlier, but they used X-radiation instead of ultraviolet light and at first concentrated more on the study of solids than on free molecules. Both ultraviolet and X-ray photoelectron spectroscopy have been extensively developed by their original discoverers and by other workers, and have found many applications in chemistry and physics. This book is an attempt to present an up-to-date view of ultraviolet photoelectron spectroscopy, and reference to X-ray work is made only when it expands upon or illuminates aspects of the valence electronic structure of molecules or ions.

1.2   MAIN FEATURES OF PHOTOELECTRON SPECTRA

In a photoelectron spectrometer, an intense beam of monochromatic (monoenergetic) ultraviolet light ionizes molecules or atoms of a gas in an ionization chamber:

$$M + hv \rightarrow M^+ + e \qquad (1.2)$$

The light used is most commonly the helium resonance line He I at 584 Å (58.4 nm), which is equivalent to 21.22 electronvolts (eV) of energy per photon. This energy is sufficient to ionize electrons from the *valence shell* of molecules or atoms, that is, from the orbitals that are involved in chemical bonding and are characterized by the highest principal quantum number of the occupied atomic orbitals. In each orbital, j, of an atom or molecule, the electrons have a characteristic binding energy, the minimum energy needed to eject them to infinity. Part of the energy of a photon is used to overcome this binding energy, I_j, and if the species is an atom the remainder, $hv - I_j$, must appear as kinetic energy (KE) of the ejected electrons:

$$KE = hv - I_j \qquad (1.3)$$

The ejected photoelectrons are separated according to their kinetic energies in an electron energy analyser, detected and recorded. The photoelectron spectrum is a record of the number of electrons detected at each energy, and a peak is found in the spectrum at each energy, $hv - I_j$, corresponding to the binding energy, I_j, of an electron in the atom, as illustrated schematically in *Figure 1.1*. If the species is a molecule, there are the additional possibilities of vibrational or rotational excitation on ionization, so the energies of the photoelectrons may be reduced:

$$KE = hv - I_j - E^*_{\text{vib., rot.}} \qquad (1.4)$$

The spectrum may now contain many vibrational *lines* for each type of electron ionized, and the system of lines that corresponds to ionization from a single molecular orbital constitutes a *band*.

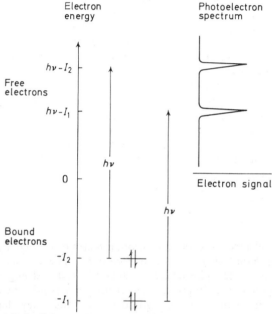

Figure 1.1. Idealized photoionization process and photoelectron spectrum of an atom

Apart from Koopmans' theorem, there are two approximate rules that make the relationship between photoelectron spectra and molecular electronic structure especially simple:

(1) Each band in the spectrum corresponds to ionization from a single molecular orbital.

(2) Each occupied molecular orbital of binding energy less than hv gives rise to a single band in the spectrum.

Because of these rules, the photoelectron spectrum is a simple reflection of the molecular orbital diagram, as illustrated in *Figure 1.1*. The rules are a simplification, however, and there are three reasons why there may, in fact, be more bands in a spectrum than there are valence orbitals in a molecule. Firstly, additional bands are sometimes found that correspond to the ionization of one electron with simultaneous excitation of a second electron to an unoccupied excited orbital. This is a two-electron process, and the bands produced in the spectrum are normally much weaker than simple ionization bands. Secondly, ionization from a degenerate occupied molecular orbital can give rise to as many bands in the spectrum as there are orbital components, because although the orbitals are degenerate in the molecule they may not be so in the positive ion. The mechanisms that remove the degeneracy are spin–orbit coupling and the Jahn–Teller effect. Thirdly, ionization from molecules such as O_2 or NO, which have unpaired electrons, can give many more bands than there are occupied orbitals in the molecule, and in such instances neither Koopmans' theorem nor the simple rules apply.

In order to introduce these main features of photoelectron spectra, it is convenient to take practical examples, starting with the spectra of atoms and proceeding to those of more complicated molecules. The spectroscopic names of atomic and molecular electronic states are constantly needed when describing the spectra, and any readers who are not familiar with them may find it helpful to consult Appendix I.

1.2.1   ATOMS

The photoelectron spectrum of atomic mercury excited by helium resonance radiation is shown in *Figure 1.2*. The vertical scale in this and all other photoelectron spectra is the strength of the electron signal, usually given in electrons per second. The absolute intensities have no physical significance because they depend on physical and experimental factors, which, although constant throughout the measurement of the spectrum, are not precisely known. The relative intensities of different peaks in the spectrum are meaningful, however, as they are equal to the relative probabilities of photoionization to different states of the positive ion, which are called the relative partial ionization cross-sections. Three horizontal scales are given in *Figure 1.2* to illustrate the relationships between measured electron energy, ionization potential and the internal excitation energy of the ions, including electronic excitation energy. Although volts (V) are the units of potential and electronvolts (eV) are units of energy, it is a usage hallowed by tradition to speak of ionization potentials as energies and to quote them in units of

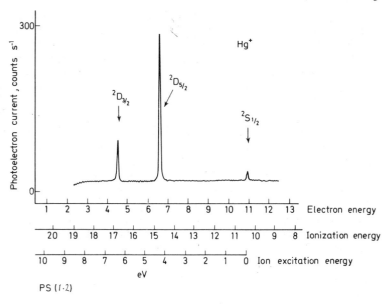

Figure 1.2. Photoelectron spectrum of mercury excited by He I (584 Å) radiation

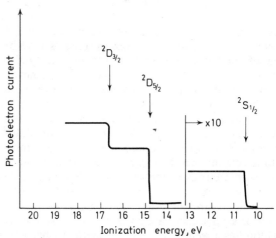

Figure 1.3. Integral photoelectron spectrum of mercury excited by He I radiation. (After Frost, D. C., McDowell, C. A., Sandhu, J. S. and Vroom, D. A., in Kendricks, E. (Editor) *Advances in Mass Spectrometry*, Vol. 4, Institute of Petroleum, London, 781 (1968))

electronvolts. Nevertheless, when energy quantities are being compared, the phrase 'ionization energy' is used frequently in this book, and henceforth the only horizontal scale given for photoelectron spectra will be one of ionization energy in electronvolts. The SI units of energy, joules, are very inconvenient in this field and are not used by spectroscopists; conversion factors for the important units are given in Appendix II.

The spectrum of mercury in *Figure 1.2* was obtained with a spectrometer in which electrons of only one energy at a time were able to reach the detector; it is called a *differential* spectrum. Spectra are sometimes encountered in *integral* form, taken with spectrometers in which all electrons of more than a certain energy can reach the detector simultaneously. The spectrum of mercury measured in such a spectrometer is shown in *Figure 1.3*, where its integral relationship to the spectrum in *Figure 1.2* is apparent. Both spectra show that Hg^+ ions are formed by photoionization in three electronic states, with ionization energies of 10.44, 14.84 and 16.71 eV. The states involved are well known from the atomic spectrum of mercury and have the designations $^2S_{\frac{1}{2}}$, $^2D_{\frac{5}{2}}$ and $^2D_{\frac{3}{2}}$, respectively. The neutral atom has the electron configuration $5d^{10}6s^2$ in the outer shells and the designation $^1S_0$. The $^2S_{\frac{1}{2}}$ state of Hg^+ is produced by the ejection of one of the 6s electrons, but both of the ^2D states are produced by the ejection of 5d electrons. A useful notation for describing these ionization processes is to write the name of the orbital from which the electron is removed with the superscript $^{-1}$. Thus the first ionization process is $6s^{-1}$ and the second and third both correspond to $5d^{-1}$. The energy difference between the $^2D_{\frac{5}{2}}$ and $^2D_{\frac{3}{2}}$ states arising from $5d^{-1}$ ionization represents a breakdown of the rule of one band per orbital, in this instance owing to spin-orbit coupling. A similar splitting is possible whenever an ionic state has both orbital and spin degeneracy. The rare gas atoms, for instance, have as their outermost orbitals completed p shells, and ionization yields p^5, a ^2P state which splits into $^2P_{\frac{3}{2}}$ and $^2P_{\frac{1}{2}}$ and gives two peaks in the spectra. Because of the breakdown of the rule of one band per orbital, Koopmans' theorem cannot be used directly to derive the orbital energy for 5d electrons in mercury or for the outer p electrons of the rare gases. A weighted mean of all the energies for ionization from a single orbital must be taken, where the weights are the statistical weights of the ionic states produced. For atoms, these are equal to $2J+1$, so the state $^2P_{\frac{3}{2}}$ of a rare gas ion has a weight of 4 and the state $^2P_{\frac{1}{2}}$ has a weight of 2. In theory, the intensities of the bands in the photoelectron spectrum should be proportional to these statistical weights, and indeed the rare gas ionizations give peaks with intensity ratios near 2:1. The

$5d^{-1}$ ionization of mercury, however, should have an intensity ratio for $^2D_{\frac{5}{2}}$ to $^2D_{\frac{3}{2}}$ of 6:4, but in fact the ratio given by the step heights in *Figure 1.3* is about 6:2.4. This is sufficient to show that the relative intensities of bands in photoelectron spectra are controlled by several more complicated factors, and these factors are discussed again in Section 1.3 and in later chapters.

1.2.2 DIATOMIC MOLECULES

The photoelectron spectrum of the diatomic molecule N_2 is shown in *Figure 1.4* as the next step in the hierarchy of complication. Three electronic states of N_2^+ are reached by photoionization with 584 Å light, and they appear in the spectrum as the sharp peak at 15.6 eV,

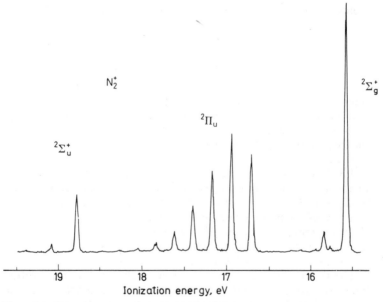

Figure 1.4. Photoelectron spectrum of nitrogen excited by He I radiation. (By courtesy of Professor W. C. Price)

the group of peaks between 16.7 and 18 eV and the weak peak at 18.8 eV. Each electronic state actually gives a group of peaks in the spectrum because of the possibility of vibrational as well as electronic excitation. Every resolved peak in the spectrum of a molecule is a single vibrational line and represents a definite number of quanta of vibrational energy in the molecular ion.

As was the case for mercury, the ionic states of N_2^+ seen in the photoelectron spectrum are well known from other forms of spectroscopy. They have the designations $X\,{}^2\Sigma_g^+$, $A\,{}^2\Pi_u$ and $B\,{}^2\Sigma_u^+$ in order of increasing ionization potential, and correspond to the ionization processes σ_g^{-1}, π_u^{-1} and σ_u^{-1}, respectively. It is clear that the bands in the spectrum for these three ionizations are very different both in the spacings of the lines within the bands, which give the sizes of the vibrational quanta in the ions, and also in the intensities of the vibrational lines. The spacings of the lines depend on the vibrational frequencies in the ions, since for the vibrational excitation energies $E_{vib.}^*$:

$$E_{vib.}^* = (v + \tfrac{1}{2})\,hv \tag{1.5}$$

Here v is the vibrational quantum number and v the frequency. The frequencies depend on the strengths of the N–N bond in the different electronic states, since for a harmonic oscillator we have

$$v = \frac{1}{2\pi}\left(\frac{k}{\mu}\right)^{\frac{1}{2}} \tag{1.6}$$

where k is the force constant and μ the reduced mass, $m_1 m_2/(m_1 + m_2)$, for a diatomic molecule. If a bonding electron is removed, the bond becomes weaker and the force constant is less in the ion than in the neutral molecule, so the vibrational frequency is lower. This is exactly what happens in the π_u^{-1} ionization of N_2, where the frequency drops from 2360 cm^{-1} in the molecule to 1800 cm^{-1} in the $A\,{}^2\Pi_u$ state of N_2^+. This reduction in frequency by a factor of 1.3 shows that the π_u electrons of nitrogen are strongly bonding in the molecule. The changes in frequency on formation of the $X\,{}^2\Sigma_g^+$ and $B\,{}^2\Sigma_u^+$ states of N_2^+ are very much smaller, and indicate a very weak bonding character. Any antibonding character of the electrons removed on ionization would be revealed by an increase in frequency, and an example of this is the π_g^{-1} ionization of molecular oxygen, shown later in *Figure 1.6*.

The relative intensities of the vibrational lines in an ionization band are also related to the bonding powers of the electron removed. Strong vibrational excitation, such as that shown in the π_u^{-1} ionization of N_2, is associated with a change in equilibrium bond length on ionization, and the relationship between them is illustrated in *Figure 1.5*. Ionization itself is a rapid process, about 10^{-15} s being required for the ejected electron to leave the immediate neighbourhood of the molecular ion. This time is so short that motions of the atomic nuclei that make up vibrations and proceed on a time scale of 10^{-13} s are effectively frozen during ionization. The internuclear distance therefore remains constant during the transition, and the process can be represented by a vertical line on a potential energy diagram. This is true of electronic transitions in general, and is known as the Franck–Condon principle.

It is a consequence of the uncertainty principle that the bond length in the molecule does not have a single precise value, but can lie within a certain range with a probability at each point given by the square of the vibrational wave-function. Vertical transitions can proceed from any point within this range, according to the instantaneous position of the nuclei at the moment of ionization.

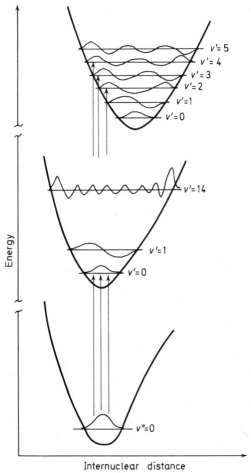

Figure. 1.5. Potential energy curves showing the form of the vibrational wave-functions, to illustrate the origin of vibrational excitation in electronic transitions

Molecules are almost all in their vibrational ground state at normal temperature, and if there is no change in bond length on ionization, the single transition from this state of the molecule to the vibrational ground state of the ion is most probable, and a single line appears in the spectrum. The reason for this effect is that the probability of each individual transition between a vibrational level in the molecule and a vibrational level in the ion is proportional to the overlap between the vibrational wave-functions in the initial and final states. The forms of the wave-functions are shown in *Figure 1.5* for different values of the vibrational quantum numbers v'' of the molecule and v' of the ion. If there is no change in bond length

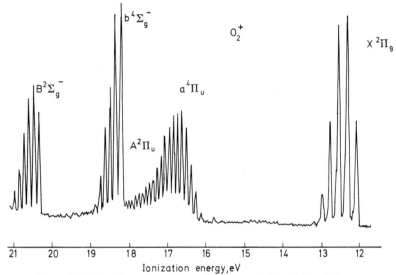

Figure 1.6. Photoelectron spectrum of oxygen excited by He I radiation

between molecule and ion, the overlap integral between the $v'' = 0$ level of the molecule and the $v' = 0$ level of the ion is large, but all overlap integrals between $v'' = 0$ and higher vibrational levels of the ion are small because positive and negative contributions cancel out. When a change in bond length occurs, the largest overlap integrals arise for transitions from the vibrationless molecule to excited vibrational levels of the ion where the wave-functions have large amplitudes at the molecular internuclear distance. Several vibrational levels of the ion can be reached and a series of lines appears in the photoelectron spectrum.

When applied to the photoelectron spectrum of nitrogen, these ideas imply that the π_u^{-1} ionization is accompanied by a change in the N–N bond length, but the σ_g^{-1} and σ_u^{-1} ionizations are not. Although the direction of the change in bond length cannot be deduced from the vibrational intensity pattern, it is clear from the decrease in vibrational frequency following π_u^{-1} ionization that the bond becomes longer. Both the change in frequency and the change in bond length on ionization reflect the character of the electron removed from the molecule, and if the electron removed is bonding, an increase in bond length and a decrease in frequency are to be expected. The photoelectron spectrum of oxygen in *Figure 1.6* illustrates these effects again, and here the first band is a π_g^{-1} ionization, the removal of an antibonding electron. There is a shortening of the O–O bond and an increase in the vibrational frequency, visible in the spectrum as strong vibrational structure with wide vibrational spacings. The second band represents the removal of an electron from the π_u orbital, which has the same character as in N_2, and the long vibrational progression and narrow spacing again reflect its bonding properties.

Oxygen is an open-shell molecule with a triplet ground state, and as a result the number of bands in the spectrum is greater than the number of orbitals. The electron configuration of the molecule is

$$\ldots 2p\sigma_g^2\, 2p\pi_u^4\, 2p\pi_g^2 \ldots {}^3\Sigma_g^-$$

The ionization of lowest energy, π_g^{-1}, gives a single state of the molecular ion, $X\,{}^2\Pi_g$, and a single band in the spectrum. The second ionization, π_u^{-1}, gives two states, a ${}^4\Pi_u$ and A ${}^2\Pi_u$, the third ionization, σ_g^{-1}, gives two states, b ${}^4\Sigma_g^-$ and B ${}^2\Sigma_g^-$, and similarly all ionizations from filled orbitals give two states, one a doublet and the other a quartet. In the photoelectron spectrum, the bands for a ${}^4\Pi_u$ and A ${}^2\Pi_u$ overlap and give the second band the appearance of being single whereas it is, in fact, double. There are five bands altogether in the photoelectron spectrum taken at 584 Å, although only three molecular orbitals are involved. In fact, the electron configuration $\pi_u^3\,\pi_g^2$ reached in the π_u^{-1} ionizations gives rise to five different electronic states, but only a ${}^4\Pi_u$ and A ${}^2\Pi_u$ can be reached by one-electron transitions. The others, ${}^2\Phi_u$ and two ${}^2\Pi_u$ states, have only a weak effect on the photoelectron spectrum[6], because the electron correlation that is necessary to make two-electron transitions intense is weak in O_2 (see Chapter 3, Section 3.2).

1.2.3 TRIATOMIC AND LARGER MOLECULES

The spectra of molecules with more than two atoms are naturally more complex, because there are generally more molecular orbitals from which ionization can take place and many different modes of vibration that may be excited on ionization. One important principle is that the vibration that corresponds most closely to the change in equilibrium molecular geometry caused by a particular ionization will be the one that is most strongly excited. In the ammonia molecule, for instance, the occupancy of the nitrogen lone-pair orbital has a strong effect on the bond angles, and when one lone-pair electron is removed the NH_3^+ ion becomes planar. The corresponding band in the photoelectron spectrum contains a long progression of lines that show excitation of the umbrella bending vibration v_2, in which the molecule can move from the pyramidal to the planar configuration. Similarly, the first band in the spectrum of acetylene (*Figure 4.5*) is a π^{-1} ionization, which weakens the C–C bond and makes it longer, and so the band contains a progression of vibrational peaks that show excitation of the C–C stretching vibration. The second and third bands, on the other hand, represent ionizations from σ orbitals with both C–C and C–H bonding character, and these bands have a more complicated vibrational structure in which two vibrational modes are involved. In a band that shows excitation of several modes, it can easily happen that the vibrational structure is so complex that the individual lines cannot be resolved and a continuous contour is seen. Such unresolved bands are, in fact, much more common than resolved bands in the spectra of all molecules with five or more atoms.

Apart from the complexity of overlapping vibrational structure, one reason for the presence of continuous bands in the spectra is a short lifetime of the ions in the molecular ionic states in which they are initially formed. If a molecular ion in a particular state has a lifetime τ before it dissociates, radiates away its excitation energy or loses its identity by internal conversion to another electronic state, each level in the original state will have an energy width ΔE, given by the uncertainty principle:

$$\Delta E \approx \frac{\hbar}{\tau} \qquad (1.7)$$

The energy uncertainty causes a broadening of all the spectral lines and may make the band continuous. This cause of broadening can occur even in the photoelectron spectra of diatomic molecules, and is considered in detail in Chapters 5 and 7.

Bands with resolved vibrational fine structure are often found in the spectra of very large molecules, despite these possible complications. They represent the ionization of electrons that are very weakly bonding, or whose bonding power corresponds closely and exclusively to a single mode of vibrational motion. Factors that are particularly favourable for the appearance of such bands are high symmetry, especially linearity or planarity, the presence of multiple bonds or rings and the presence of heteroatoms with non-bonding lone-pair electrons. Vibrational fine structure is most often to be found in the first band at the lowest ionization potential in the spectrum of a complex molecule because the outermost electrons are least likely to have strong bonding character.

1.3 INTENSITIES OF PHOTOELECTRON BANDS

We have so far considered the energies of photoelectron bands and lines and the intensities of lines within the vibrational structure of a band. The intensities of the bands themselves also give useful information, which is needed in order to be able to analyse the photoelectron spectra of molecules whose ionic states have not already been identified. The areas of photoelectron bands in a spectrum are approximately proportional to the relative probabilities of ionization to the different ionic states, but experimental factors are also involved. Practical aspects of this problem are discussed in Chapter 2 ; suffice it to say that the relative ionization probabilities, called relative partial ionization cross-sections, can be derived from the measured spectra, and it is to these derived quantities that the following considerations apply. For resolved bands, the intensity is obtained by summing the intensities of all the vibrational lines.

1.3.1 LIMITING RULES FOR RELATIVE INTENSITIES

The relative partial cross-sections for photoionization to particular ionic states are sometimes also called the branching ratios in ionization, and the most important factors that determine them can be expressed in three approximate rules, as follows.

Rule 1. The partial cross-section for ionization from a given orbital is proportional to the number of equivalent electrons that are available to be ionized. Ionization from an orbital that contains four electrons is twice as probable as ionization from an orbital that contains only two electrons, and all ionizations from orbitals that contain two electrons should give bands of the same intensity. This rule is most nearly valid when the comparison is made between ionizations from molecular orbitals that are made up from the same set of atomic orbitals, but despite this restriction it is extremely useful as it is the only rule needed when interpreting the band intensities in the spectra of closed-shell molecules. As an example of the application of this rule, part of the photoelectron spectrum of biphenyl is shown in *Figure 1.7*, where it can be seen that the relative intensities of the first three bands, allowing for some overlap,

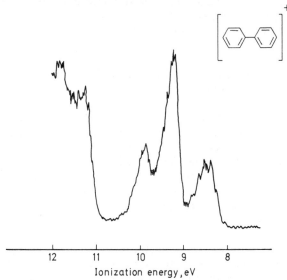

Figure 1.7. Partial photoelectron spectrum of biphenyl, showing the π electron ionization region

are approximately $1:2:1$. There is evidence from the spectra of other aromatic compounds that the low ionization potential region includes only π electron ionization bands, and the spectrum can therefore be compared with the π electron configuration expected on the basis of Hückel molecular orbital theory. The predicted π orbital pattern for biphenyl is

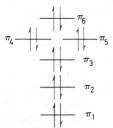

Of the outer π orbitals, π_4 and π_5 are degenerate and form a degenerate orbital that contains four electrons; the intensity pattern seen in

the photoelectron spectrum corresponds exactly with the π orbital pattern. As a second example, the photoelectron spectrum of carbon tetrachloride is shown in *Figure 1.8*. Molecular orbital theory indicates that the outermost occupied orbitals in this molecule are made up from the non-bonding chlorine p atomic orbitals. In tetrahedral symmetry, these orbitals combine to form the molecular

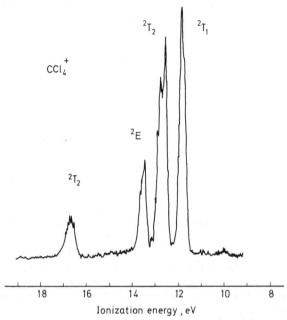

Figure 1.8. Photoelectron spectrum of carbon tetrachloride

orbitals t_1, t_2 and e, of which the first two are triply degenerate and the third doubly degenerate. The relative areas of the first three bands in the spectrum are $1.9:2.6:1.0$; this is sufficient to give a strong indication that the e^{-1} ionization is the third band, although the relative areas are rather far from the predicted ratio of $3:3:2$. Comparisons with other spectra and with calculations confirm this assignment, and show that the bands represent t_1^{-1}, t_2^{-1} and e^{-1} ionizations, respectively.

Rule 2. The partial ionization cross-section is proportional to the statistical weight of the ionic state produced. In ionization from closed-shell molecules, this rule is equivalent to Rule 1, but for the ionization of atoms and open-shell molecules it is more extensive. The $^2P_{\frac{3}{2}}$ and $^2P_{\frac{1}{2}}$ states of the rare gas atomic ions arise from ionization from the same orbital but have statistical weights of 4 and 2, respectively, and the two peaks in the spectrum have an intensity ratio of 2:1. The ionization of molecular oxygen produces ions in both quartet and doublet states, and the bands due to the quartet states should be twice as intense as those for doublet states that arise from the same orbital ionization. Both Rules 1 and 2 apply to ionization from the closed shells of an open-shell molecule such as O_2. The total cross-section for ionization from a filled orbital, measured as the summed intensities of all the bands that correspond to the resulting electronic configuration of the ion, should be proportional to the number of equivalent electrons in that orbital. The relative intensities of the bands that correspond to a single electron configuration of the ions should be proportional to the statistical weights of the different ionic states, that is, their spin–orbital degeneracies. Nitric oxide, for example, has the electron configuration

$$\ldots 4\sigma^2 \, 5\sigma^2 \, 1\pi^4 \, 2\pi \ldots \, ^2\Pi$$

The nitric oxide ion produced by $1\pi^{-1}$ ionization has the configuration $4\sigma^2 \, 5\sigma^2 \, 1\pi^3 \, 2\pi$, and this configuration gives the six ionic states $^3\Sigma^+$, $^3\Delta$, $^3\Sigma^-$, $^1\Sigma^-$, $^1\Delta$ and $^1\Sigma^+$, all of which are seen in the photoelectron spectrum[7]. The relative intensities of the photoelectron bands are approximately in agreement with the statistical weights of 3, 6, 3, 1, 2 and 1, respectively. Similarly, the ionization $5\sigma^{-1}$ gives $^1\Sigma$ and $^3\Sigma$ states with statistical weights, and also relative intensities in the spectrum, of 1:3. According to Rule 1, the total intensity of all the $1\pi^{-1}$ ionization bands together should be double that of the summed $5\sigma^{-1}$ bands, but this is not borne out in the spectrum.

Rule 3. In ionization from the open shell itself of an open-shell molecule, the relative band intensities are, in general, proportional to the coefficients of fractional parentage[8]. This rule is due to Cox and Orchard[9], who also first enunciated Rules 1 and 2 as they apply to open-shell molecules. In instances when the molecular configuration generates only one term, Rule 3 is equivalent to Rule 2, and this is so if the open shell contains either a single electron or one electron less than the number required for a complete shell. For the ionization of molecules that have more than one open shell, a further and more complex rule has been given[10].

1.3.2   BAND INTENSITY AND ORBITAL CHARACTER

The three limiting rules given above are based purely on statistical considerations; they take no account of any differences in ionization cross-section that are dependent on the character of the orbitals. Relative band intensities in photoelectron spectroscopy often do depend on the nature of the orbitals, and their size, number of nodes and their localization in the molecule are probably the main factors. In the photoelectron spectrum of atomic mercury, the $6s^{-1}$ and $5d^{-1}$ ionizations have an intensity ratio of $1:21$ (*Figure 1.3*), whereas the orbital occupancies predict a ratio of $1:5$. Fortunately, the deviations are not often so large, and the partial cross-sections for ionizations from molecular orbitals that are made up principally from the same atomic orbitals are usually more nearly in accordance with the limiting rules. This includes many of the most important practical instances, and when deviations occur they can also be useful as they may indicate participation by a different atomic orbital in the formation of the molecular orbital concerned. There are no hard and fast rules for atomic orbital characteristic cross-sections in ultraviolet photoelectron spectroscopy, but experience allows some generalizations to be made. Firstly, in photoionization by 584 Å light, the cross-sections for ionization from the valence orbitals are higher in heavy atoms than in light ones, and there is a large increase in cross-section between atoms of the first and second (full) rows of the Periodic Table. Thus the outer p orbitals of sulphur have a higher cross-section than those of oxygen, the chlorine p electrons have a higher cross-section than those of fluorine, and both sulphur and chlorine p^{-1} ionizations give stronger bands than carbon p^{-1} ionization. This trend is at least partially reversed, however, when light of shorter wavelength, particularly the 304 Å line of helium, is used for ionization. A second generalization is that as the Periodic Table is crossed from left to right, the cross-sections for ionization of atomic s-type orbitals decrease relative to those for p-type valence orbitals in the same atoms[11].

When light of very short wavelength is used for ionization, as in X-ray photoelectron spectroscopy, the characteristic atomic orbital cross-sections are very different from those encountered in ultraviolet photoelectron spectroscopy, and moreover they can be related quantitatively to band intensities in the spectra. The molecular orbitals are modelled as linear combinations of atomic orbitals, and a characteristic molecular orbital cross-section is obtained by summing characteristic atomic orbital cross-sections multiplied by the squares of the atomic orbital coefficients. This model has given very good agreement between calculated partial cross-sections and the experimental spectra[12] and should be a most useful tool in the study of molecular electronic structure. For photoelectron spectra excited by ultraviolet light, a related but more complicated model has been proposed by Thiel and Schweig[13], who have tested it on the spectra of some linear molecules. The agreement found is encouraging, and it seems that the comparison of measured and calculated band intensities may become a reliable aid in the analysis of photoelectron spectra.

1.4   THE ANALYSIS OF PHOTOELECTRON SPECTRA

The analysis of a photoelectron spectrum consists in the assignment of each band in the spectrum to a particular electronic state of the molecular ion and to ionization from a particular orbital of the molecule. The first step in this process is to ascertain from the molecular orbital model those orbitals which are occupied in the molecule, the characters of these orbitals, whether they are bonding, antibonding or non-bonding and degenerate or non-degenerate. The order of the orbitals and ionization bands in terms of energy is seldom reliably predicted by a qualitative molecular orbital model, although some clues are always available; the main task in the analysis is to determine this order. The characters of the orbitals are shown in the photoelectron spectrum by the vibrational structures and intensities of the bands, and these are the first points to be compared with the theoretical model. It is also very helpful to make comparisons with the photoelectron spectra of related molecules, whether the spectra have been analysed or not, in order to establish if there are common features or regular trends that can be of help with the assignments. Finally, there is sometimes evidence from outside photoelectron spectroscopy that can assist with the assignments, generally from other forms of spectroscopy. In the next section some examples of the analysis of photoelectron spectra on the basis of internal evidence are presented, and sources of external evidence are afterwards discussed in the final section of this chapter. Many more examples of the analysis of photoelectron spectra can be studied in the book on photoelectron spectroscopy by Turner and his collaborators[14].

1.4.1   EXAMPLES

Mercury(II) chloride

Mercury(II) chloride is a linear molecule that contains 16 valence electrons of which four are in chlorine 3s orbitals with too high a binding energy to be ionized by 21.22 eV photons.* According to the molecular orbital model, the remaining 12 electrons are accommodated as follows:

$$\sigma_g^2 \; \sigma_u^2 \; \pi_u^4 \; \pi_g^4 \ldots {}^1\Sigma_g^+$$

The σ_g orbital is made up from two chlorine 3p orbitals bonding with the 6s orbital of mercury, and is expected to be the main bonding orbital. The π_g orbital, on the other hand, is an out-of-phase combination of chlorine 3p orbitals of π symmetry, and is purely non-bonding in the absence of d orbital participation. A strongly bonding orbital is likely to be deeper lying than a non-bonding orbital, so it is safe to predict from the model that σ_g^{-1} requires a higher ionization energy than π_g^{-1}. The bonding properties of σ_u and π_u are more problematical, as they involve the participation of the unoccupied mercury 6p orbitals, which are of high energy: no prediction of the relative energies of the σ_u^{-1} and π_u^{-1} ionizations can be made.

The photoelectron spectrum of mercury(II) chloride in the valence region is shown in *Figure 1.9*, and contains four bands with relative areas of $2.1:2.0:1.1:0.3$ in order of increasing ionization potential, which match the four valence orbitals in number. According to the limiting rules for intensities, the first two bands must be the two π^{-1} ionizations from degenerate orbitals and the third and fourth must be σ^{-1} ionizations. The fourth band probably represents ionization from an orbital with strong Hg character because of its deviant low intensity, and this can only be σ_g. The first band in the spectrum consists of two sharp peaks that show the non-bonding character of the orbital ionized, and of the two π orbitals this must be π_g. The two peaks arise from spin–orbit splitting of the ${}^2\Pi_g$ state into ${}^2\Pi_{\frac{3}{2}g}$ and ${}^2\Pi_{\frac{1}{2}g}$, and they are of equal intensity as the two states have equal statistical weights. This is because in a linear molecule angular momentum is quantized only along the molecular axis and all Π, Δ or Φ states have an orbital degeneracy of two, irrespective of the value of Λ. The π_u orbital is an in-phase combination of chlorine 3p with mercury 6p orbitals and is bonding; the breadth of the second band in the photoelectron spectrum is in accordance with this character. The ${}^2\Pi_u$ state must also be split by spin–orbit coupling, but the splitting is obscured by the unresolved vibrational structure.

With mercury(II) chloride, information from band intensities and band shapes, together with qualitative considerations of the form

---

* The $5d^{10}$ electrons of the mercury atom can also be ionized with 21.22 eV photons, but as analysis of the $5d^{-1}$ bands involves some complications it is dealt with later in Chapter 4.

of the molecular orbitals, is sufficient for an analysis of the spectrum to be made. That the bands represent π_g^{-1}, π_u^{-1}, σ_u^{-1} and σ_g^{-1} ionizations, in order of increasing ionization potential, is confirmed

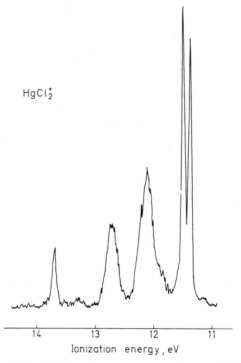

Figure 1.9. Partial photoelectron spectrum of mercury(II) chloride

by comparison with the spectra of the other mercury(II) halides and the other 16 electron linear triatomic molecules CO_2 and COS. The four states of $HgCl_2^+$ can be labelled $X\,^2\Pi_g$, $A\,^2\Pi_u$, $B\,^2\Sigma_u^+$ and $C\,^2\Sigma_g^+$, and the task of analysis is complete.

Water

The H_2O molecule is bent and has only C_{2v} symmetry, so that it has no degenerate orbitals. The analysis of its photoelectron spectrum (*Figure 1.10*) can be made on the basis of the vibrational structure of the different bands. The molecular orbitals of H_2O with binding energies less than 21.22 eV are essentially formed from the three p orbitals of the oxygen molecule. One p orbital rises out of the molecular plane; it has b_1 symmetry in C_{2v} and is completely non-bonding. The next p orbital is in the plane and bisects the HOH bond angle; it has a_1 symmetry and, although it is weakly O–H bonding, its most important characteristic is that its occupancy determines the HOH bond angle. The third p orbital is in the plane and perpendicular to the other two orbitals; with the

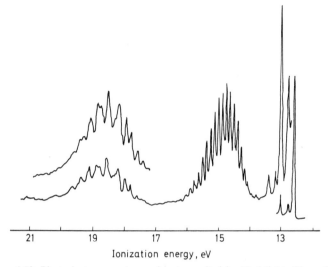

Ionization energy, eV

Figure 1.10. Photoelectron spectrum of water excited by He I light. (From Potts, A. W. and Price, W. C., *Proc. R. Soc., Lond.*, **A326**, 181 (1972), by courtesy of the Council of the Royal Society)

hydrogen 1s orbitals, it forms the main O–H bond. The first band in the photoelectron spectrum, with its single sharp peak and weak vibrational structure, is clearly ionization from the non-bonding orbital b_1. The second band consists of a very long progression in which the first few vibrational intervals are about 900 cm^{-1}. The O–H stretching vibrations, v_1 and v_3, have frequencies so much higher than this that the mode excited can only be the bending vibration, v_2, whose frequency in the neutral molecule is 1595 cm^{-1}. A triatomic molecule has only three vibrational modes, so there is no doubt about this vibrational assignment. The band assignment follows at once, as only the a_1 orbital has the angle-determining character that could cause such strong excitation of v_2 on ionization. It is found, in fact, that in the $^2A_1$ state which results, H_2O^+ is linear in the equilibrium position. The third band in the spectrum must be assigned to ionization from the remaining orbital, b_2, and its vibrational structure is in accordance with this. It shows excitation of the O–H stretching vibration, v_1, probably with v_2 also excited. The reduction in frequency of v_1 on ionization and the shape of the band both indicate the ionization of a strongly bonding electron. The interpretation of the vibrational structure of the bands therefore leads to the identification of the three states of H_2O^+ as $^2B_1$, $^2A_1$ and $^2B_2$, in order of increasing ionization potential.

Methanol

In the photoelectron spectrum of methanol shown in *Figure 1.11*, there is little resolved vibrational structure, and analysis of the spectrum depends mainly on comparison with the photoelectron spectrum of water. The molecule has at the most a plane of symmetry,

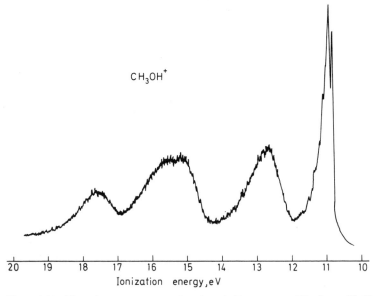

CH_3OH^+

Ionization energy, eV

Figure 1.11. Photoelectron spectrum of methanol. (By courtesy of Professor W. C. Price)

so the symmetry labels of the different molecular orbitals are not very informative. It is more useful to describe the orbitals approximately in terms of their atomic orbital parentage and bonding character, which can be derived either by calculation or from empirical considerations. The occupied orbitals of the valence shell are:

3a′	O 2s	Weakly bonding
4a′	C 2s	Weakly bonding
5a′	O 2p + C 2p + H 1s	HOC bonding
1a″ } 6a′ }	C 2p + H 1s	CH_3 bonding
7a′	O 2p	HOC angle-determining
2a″	O 2p	Out-of-plane, non-bonding

Of these seven orbitals, the 3a′ oxygen 2s orbital and the 4a′ carbon 2s orbital have too high a binding energy to be observed with He I excitation, so five bands are to be expected. The fact that only four distinct bands appear shows that two must be overlapping, and the most likely pair is 1a″ and 6a′. If the OH group of methanol lay along the axis of the CH_3 group, 1a″ and 6a′ would be degenerate and form an e orbital responsible for C–H_3 bonding, and the splitting between the two orbitals in methanol is therefore probably small. Only the third band in the spectrum is intense enough to represent ionization from two occupied orbitals, so it can be tentatively assigned to 1a″ and 6a′ together. This assignment is supported by the fact that similar bands of double intensity and attributed to the CH_3 e orbitals are found at the same ionization potential in the spectra of CH_3I, CH_3CN and CH_3SH. The three bands that remain unassigned are the first, third and fourth, and the orbitals whose ionization they must represent, 2a″, 7a′ and 5a′, are all based on oxygen 2p orbitals and correspond to the three occupied orbitals of water. The first band of methanol is identified by its narrow contour and vibrational structure as the $2a''^{-1}$ ionization, the second band must be ionization from the angle-determining 7a′ orbital and the fourth band is probably $5a'^{-1}$ ionization from the main H–O–C bonding orbital. This analysis of the spectrum is consistent with all the evidence from the spectrum itself and from the spectra of related compounds, but because of the lack of vibrational structure in the spectrum it is still not completely certain. In instances such as this, it is useful to seek external evidence, as discussed in the next section.

1.4.2 EXTERNAL EVIDENCE FOR BAND ASSIGNMENTS

An important aid to the analysis of photoelectron spectra is comparison with detailed molecular orbital calculations of orbital energies and ionization energies. This topic will be described separately in Chapter 4, and here the experimental evidence relevant to the analysis of photoelectron spectra is considered. The most useful evidence is spectroscopic, and a distinction can be made between information about the molecular ground states and information about the ionic states themselves.

Most molecules have singlet ground states in which the electronic wave-function is totally symmetric, but this is seldom true of atoms, and there are also many molecules that have open shells, particularly among transition metal compounds. Molecular oxygen is the most familiar example, and if its ground state were $^1\Delta_g$ instead of $^3\Sigma_g^-$, its photoelectron spectrum would be very different. If a molecule has an open shell, the species of the ground state may be determined by optical spectroscopy or electron spin resonance spectroscopy or from magnetic susceptibility measurements. Evidence from electron spin resonance spectroscopy, for instance, has been used in the analysis of the spectra of transition metal π-arene complexes[15]. If the ground state of an atom or molecule happens to be a spin doublet, its species identifies the outermost orbital directly. Nitric oxide has a $^2\Pi$ ground state, so the outermost occupied orbital is a π orbital and the first band in the photoelectron spectrum represents a π^{-1} ionization. The ground states of NF_2 and ClO_2 both have the species $^2B_1$, and the first bands in their spectra are therefore attributed to b_1^{-1} ionizations[16, 17].

The electronic excited states of molecules that are studied in optical spectroscopy are reached by the transfer of one electron from an occupied to an unoccupied orbital. Two types of transitions are distinguished, *intervalence* or *sub-Rydberg* transitions and *Rydberg* transitions. In intervalence transitions, the electron is promoted to an orbital made up from atomic orbitals with the same principal quantum number as those which make up the valence shell. The excited states produced are the lowest-lying excited states of the molecule, and are seen as absorption or emission bands in the visible and ultraviolet regions of the spectrum. The characteristics of the absorption bands can sometimes show which orbital of the molecule is outermost, but deductions about the molecular orbital diagram are difficult to make because two orbitals of unknown energy are involved in every transition. In Rydberg transitions, the excited state has one electron in an orbital with a principal quantum number greater than that characteristic of the valence shell. The excited electron is in a Rydberg orbital, which typically has a large radius and high principal quantum number and approaches a hydrogenic atomic orbital in form. The electron moves in a field defined by the positively charged core, which is the molecular ion in a particular electronic state. Rydberg states with successively higher principal quantum number form a series in the spectrum that converges on the ionization limit at which the excited electron is free, and the molecular ion which formed the core remains. The frequencies at which absorption to Rydberg states occurs can be expressed by the equation

$$\nu = \nu_\infty - \frac{R}{(n-\delta)^2} \qquad (1.8)$$

where ν_∞ is the ionization limit, R the Rydberg constant, n the principal quantum number and δ the quantum defect or Rydberg correction. This equation is exactly the same as that for atomic Rydberg bands. For molecules, the Rydberg bands have vibrational and rotational structure, but as the outer electron is so far away it has little influence on the bonding in the core, and the vibrational structures are very similar to the vibrational structures seen in the photoelectron spectrum. This fact can be used to help in the analysis of the Rydberg bands; each member of a series has nearly the same vibrational structure, and it is the same as the structure of the photoelectron band at the ionization potential on which the series converges. If the photoelectron spectrum of a molecule contains only continuous bands, there may be no resolved Rydberg bands and analysis is impossible. Even when the bands are well resolved, the spectrum of Rydberg states is often very complex because there are usually several series converging on each ionization limit. The Rydberg bands are found in the far ultraviolet and vacuum ultraviolet regions of the spectrum, where the experimental difficulties of optical spectroscopy are severe, but the new experimental technique of electron energy loss spectroscopy (Chapter 3, Section 3.5.3) makes it much easier to observe them. They are generally found within an energy range of about 2 eV from the ionization limit on which their series converges.

The importance of the study of Rydberg states for the analysis of photoelectron spectra is that from the quantum defects and the number of series that converge on a given limit, the symmetry of the molecular orbital from which ionization takes place can be deduced. Each Rydberg state is characterized by its electron configuration and symmetry species; for water, for instance, one such state is

$$1b_2^2\ 1a_1^2\ 1b_1\ n{\rm sa}_1 \ldots\ ^1B_1$$

The Rydberg orbital is given last in the electron configuration with a notation that shows its principal quantum number, n, its s, p or d character and its symmetry species in the molecular point group, here a_1. The over-all symmetry species of the state is indicated in the usual way, and can be obtained by taking the direct product of the species of the Rydberg orbital with that of the molecular ionic

26 PRINCIPLES OF PHOTOELECTRON SPECTROSCOPY

core. In C_{2v} symmetry, an s orbital always has a_1 symmetry, but a p orbital can be a_1, b_1 or b_2 and a d orbital can be a_1 (twice), a_2, b_1 or b_2. These are the species obtained by resolution of the atomic species into point groups of lower symmetry[18]. For removal of a $1b_1$ lone-pair electron from water, six separate Rydberg states with p-type Rydberg orbitals can be constructed:

$$1b_2^2\, 3a_1^2\, 1b_1\, npa_1 \ldots {}^1B_1 \text{ or } {}^3B_1$$
$$1b_2^2\, 3a_1^2\, 1b_1\, npb_1 \ldots {}^1A_1 \text{ or } {}^2A_1$$
$$1b_2^2\, 3a_1^2\, 1b_1\, npb_2 \ldots {}^1A_2 \text{ or } {}^3A_2$$

Now, the ground state of water is a $^1A_1$ state, and the normal optical selection rules[19] exclude the formation by electric dipole transitions of all the triplet states and also the $^1A_2$ state. Therefore, only two Rydberg series with p character that converge on the limit for a b_1^{-1} ionization in a C_{2v} molecule are to be expected. The s, p or d character of the observed series can be discovered from the quantum defects, as these defects are characteristic of the penetration of the Rydberg orbitals into the core. For ns electrons, the defect is usually 0.9 to 1.2, for np electrons 0.3 to 0.5 and for nd electrons less than 0.1, all of these values being appropriate for small molecules of light elements; excitations to f or higher orbitals are very rare.

In the spectrum of water, two series with defects characteristic of p character are indeed found, leading to the first ionization potential, the $1b_1^{-1}$ ionization. When the above analysis is repeated for ionizations from the $3a_1$ and $1b_2$ orbitals, it is found that three and two p-type Rydberg series, respectively, would be expected to converge on their ionization limits. Because of the broad bonding contours of the $3a_1^{-1}$ and $1b_2^{-1}$ ionization bands, these series have not been found, but the observation of two p series converging on the first limit proves that the outermost orbital of water is of b_1 or b_2 symmetry. The principle bears repeating that the analysis depends on the *number* of series of each type, s, p or d, that converge on a particular limit. Some other examples are given in Chapter 4 in connection with analysis of the photoelectron spectrum of benzene, where it is important, for instance, that in a_{2u}^{-1} ionization no Rydberg excitations of p character are allowed. The symmetry tables needed for the analysis of Rydberg series are to be found in Herzberg's book[18].

REFERENCES

1. AL-JOBOURY, M. I. and TURNER, D. W., *J. chem. Soc.*, 5141 (1963)
2. AL-JOBOURY, M. I. and TURNER, D. W., *J. chem. Phys.*, 37, 3007 (1962)

3. VILESOV, F. I., KURBATOV, B. L. and TERENIN, A. N., *Dokl. Akad. Nauk SSSR*, **138**, 1329 (1961) [in Russian; English translation in *Sov. Phys. Dokl.*, **6**, 490 (1961)]
4. SIEGBAHN, K., NORDLING, C., FAHLMAN, A., NORDBERG, R., HAMRINN, K., HEDMAN, J., KOHANSSON, G., BERGMARK, T., KARLSSON, S. E., LINDGREN, I. and LINDBERG, B., *Nova Acta Regiae Soc. Sci. Upsal., Ser. IV*, **20** (1967)
5. KOOPMANS, T., *Physica*, **1**, 104 (1933)
6. EDQVIST, O., LINDHOLM, E., SELIN, L. E. and ASBRINK, L., *Physica Scripta*, **1**, 25 (1970)
7. EDQVIST, O., LINDHOLM, E., SELIN, L. E., SJOGREN, H. and ASBRINK, L., *Ark. Fysik*, **40**, 439 (1970)
8. BRINK, D. M. and SATCHER, G. R. *Angular Momentum*, Oxford University Press, London (1968)
9. COX, P. A. and ORCHARD, F. A., *Chem. Phys. Lett.*, **7**, 273 (1970)
10. COX, P. A., EVANS, S. and ORCHARD, F. A., *Chem. Phys. Lett.*, **13**, 386 (1972)
11. PRICE, W. C., POTTS, A. E. and STREETS, D. G., in Shirley, D. A. (Editor) *Electron Spectroscopy*, North Holland, Amsterdam, 187 (1972)
12. GELIUS, U., in Shirley, D. A. (Editor) *Electron Spectroscopy*, North Holland, Amsterdam, 311 (1972)
13. THIEL, W. and SCHWEIG, A., *Chem. Phys. Lett.*, **12**, 49 (1971); **16**, 409 (1972)
14. TURNER, D. W., BAKER, A. D., BAKER, C. and BRUNDLE, C. R., *Molecular Photoelectron Spectroscopy*, Wiley, New York (1970)
15. EVANS, S., GREEN, J. C. and JACKSON, S. E., *J. chem. Soc., Faraday Trans. II*, **68**, 249 (1972)
16. FROST, D. C. and MCDOWELL, C. A., *J. chem. Phys.*, **54**, 1872 (1971)
17. CORNFORD, A. B., FROST, D. C., HERRING, F. G. and MCDOWELL, C. A., *Chem. Phys. Lett.*, **10**, 345 (1971)
18. HERZBERG, G., *Electronic Spectra and Electronic Structures of Polyatomic Molecules*, Van Nostrand, Princeton, N.J., 576–577 (1966)
19. HERZBERG, G., *Electronic Spectra and Electronic Structures of Polyatomic Molecules*, Van Nostrand, Princeton, N.J., 132 (1966)

2 Experimental Methods

2.1 INTRODUCTION

The essential components of a photoelectron spectrometer are a lamp that produces suitable radiation, an ionization chamber in which molecules can be ionized at a defined electrical potential, an electron energy analyser, an electron detector and a recorder. These components are shown schematically in *Figure 2.1* and are discussed

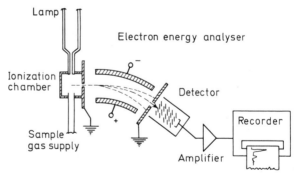

Figure 2.1. Essentials of a photoelectron spectrometer. All the electron optics must be within a vessel evacuated to 10^{-5} mm Hg or less

sequentially in the next section. A photoelectron spectrum is measured by varying the energy of the photoelectrons allowed to reach the detector and recording the rate at which electrons of each energy arrive. An ideal photoelectron spectrometer should give a complete and accurate record of the energy distribution of the electrons that are actually emitted from the sample molecules, but no practical instrument does this. The performance of real instruments can be discussed in terms of the accuracy with which the energies and relative intensities characteristic of the true distribution can be measured.

An important factor that affects the accuracy of energy measurements is the resolution of a photoelectron spectrometer, the smallest energy difference between two groups of electrons that will result in separate photoelectron peaks being registered. The best resolution so far achieved is 4–5 meV, which corresponds to 32–42 cm^{-1}, so photoelectron spectroscopy still has some way to go before it can compete in resolution with optical spectroscopy, where a resolution of 0.1 cm^{-1} is normal. Accurate energy measurements also depend on the proper calibration of the instrumental energy scale, which is more difficult than in optical spectroscopy because of the susceptibility of electrons to stray electric and magnetic fields.

The precision of intensity measurements in photoelectron spectroscopy is largely a matter of the time allowed for the experiments. Each point in a photoelectron spectrum represents the detection of a definite number of electrons, the product of the electron arrival rate and the experimental time, and the intensity has an uncertainty equal to the square root of this number. If 100 electrons are recorded, the statistical uncertainty (one standard deviation) is ± 10 electrons, or $\pm 10\%$, but if the experimental time is quadrupled and 400 electrons are recorded, the uncertainty is only ± 20 electrons, or $\pm 5\%$ of the signal. The result of each single measurement is a random quantity, like the number of radioactive decays recorded by a Geiger counter. This is the origin of the random fluctuations in photoelectron spectra, which are called statistical noise. The noise can be reduced in comparison with the signal only by increasing the number of electrons recorded at each point in the spectrum. A high-resolution spectrum must contain more individual points than a low-resolution spectrum, and therefore requires a longer experimental time for equal precision in the intensities to be attained. A vital object in the design of photoelectron spectrometers is to ensure that the arrival rates of electrons are high enough to make the measurement of complete spectra possible in a reasonable time.

2.2 LIGHT SOURCES

The most desirable properties of a light source for photoelectron spectroscopy are narrowness of the principal ionizing line, the absence of any other lines, and high intensity. Much development effort on commercial photoelectron spectrometers has been concentrated on the light sources because deficiencies there are hard to compensate for in the remainder of the instrument, whereas a high intensity of the light source can permit the use of high resolution without making the measurement of a spectrum unduly lengthy.

2.2.1 DISCHARGES IN HELIUM AND OTHER GASES

The most useful and widely used light source in photoelectron spectroscopy is a discharge in pure helium, which gives the He I resonance line at 584 Å, equivalent to a photon energy of 21.22 eV, as its main output. This light is energetic enough to cause ionization of the majority of valence electrons, and is not accompanied by any lines of lower energy down to about 4 eV. The development of the He I resonance lamp as a source of ionizing radiation was a major factor in the early success of photoelectron spectroscopy[1]. In different forms of the lamp, radiation is excited by a high-voltage direct-current discharge in a capillary, by microwave discharge or by a high-current arc discharge using a heated cathode as electron source. Some designs of these three types of lamp are shown in *Figure 2.2*. The transition responsible for producing the 584 Å line is from He $1s2p \ldots {}^1P$ to the ground state $1s^2 \ldots {}^1S$; higher members of this series, that is, $1s \, np \ldots {}^1P$ to the ground state, are also present in the output of the lamps, but their intensity is not more than a few per cent of that of the 584 Å line. The lines of this series, starting from the 584 Å line, are called He $I\alpha$, He $I\beta$, etc., and are listed in *Table 2.1*; when the higher lines are not referred to specifically, the Greek letters can be omitted and the designation He I alone in this book and elsewhere always refers to the 584 Å line.

All the transitions of the He I series are fully allowed in absorption as well as emission, and for this reason the 584 Å and other lines are always self-reversed. The hot gas in the discharge emits lines that are broadened by the Doppler effect of the velocities of the emitting atoms, but it passes through cold gas outside the discharge region, which absorbs the centres of the lines completely, so that the effective widths of the lines are then much greater than their intrinsic widths. Samson[2] has examined the sharpness and intensities of the He I lines produced by some practical photoelectron spectrometer lamps and found that their energy widths were of the order of a few millielectronvolts and therefore small enough not to degrade spectrometer resolutions. Nevertheless, it is important to minimize the amount of cold helium gas through which the light has to pass if

31

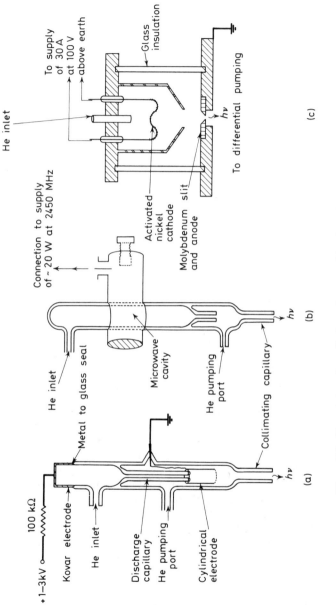

Figure 2.2. Three types of gas discharge lamp used in photoelectron spectroscopy: (a) a simple all-glass capillary discharge; (b) microwave discharge; (c) high-current arc discharge. The working helium pressure is 0.1–1 mm Hg in all instances

this small width is to be attained, and also to avoid loss of intensity. The wavelengths and energies of the He I lines are given in *Table 2.1*, which also shows the commonest impurity lines found in the output of helium discharge lamps.

Table 2.1   LINES FROM HELIUM DISCHARGE LAMPS

Line	Wavelength, Å	Energy, eV	Intensity
He Iα	584.3340	21.2175	100
He Iβ	537.0296	23.0865	2
He Iγ	522.2128	23.7415	0.5
He IIα	303.781	40.8136	< 1
He IIβ	256.317	48.3702	
He IIγ	243.027	51.0153	
He IIδ	237.331	52.2397	
H Lyman α	1215.67	10.1986	
N I	1134.4147	10.9290	
Ne Iα	735.895	16.8476	

The relative intensities given are typical for a capillary discharge under normal conditions for He Iα output. The relative abundance of the lines of higher energy can be increased at low pressures, as explained in the text. The last three lines mentioned arise from common impurities either in the helium supply or de-gassed from the lamp structure.

Discharges in helium can also generate a series of lines from ionized helium, He II, with the main line at 303 Å, equivalent to 40.81 eV. This line is of great interest for photoelectron spectroscopy as it makes complete valence shells accessible. In order to produce substantial amounts of He II light from a discharge lamp, it is only necessary to increase the current density and to decrease the helium gas pressure. When this is first attempted, the immediate result is that either the discharge goes out, or large numbers of electrons as well as photons are produced. The first problem can be overcome to some extent by increasing the area of the cathode and by providing a high-voltage power supply (about 10 kV), and the second can be countered by providing electric or magnetic deflection of the charged particles before they reach the ionization source. The output of the lamp may then contain a substantial proportion of He II light, perhaps as much as He I, although its total intensity will be much reduced. The production of He II is even more sensitive than that of He I to the presence of foreign gases in the lamp, and for this reason the choice of materials for lamp construction is important. Many early lamps contained discharge capillaries made of Pyrex or fused quartz, but recently the best grade (grade 'M') of boron nitride together with pure tantalum electrodes have been used increasingly. He II radiation free from He I can be obtained without a monochromator by total absorption of He I in a gas cell or in a thin foil. This relies on the very much higher absorption by compounds of light atoms at 584 Å than at 304 Å. Lamps that operate on these principles and provide almost pure He II radiation have been designed[3] and their use in photoelectron spectroscopy is just beginning.

The gas discharge lamps can be run on gases other than helium or on gases mixed with helium to produce ionizing lines of lower energy. The most intense is hydrogen Lyman α, the most useful the pair of neon resonance lines. All of these sources contain ionizing radiation of several frequencies and different intensities, which complicate the interpretation of the observed spectra. Details of the energies of the useful lines, together with some very approximate relative intensities, are given in *Table 2.2*.

Table 2.2   LINES FROM DISCHARGES IN GASES OTHER THAN HELIUM

Gas	Line	Energy, eV	Intensity
Neon	Ne Iα	16.6704	15
		16.8476	100
	Ne Iβ	19.6877	<1
		19.7792	<1
	Ne II	26.8132	(100)
		26.9100	(100)
	Ne II	27.6858	(20)
		27.7616	(20)
		27.7827	(20)
		27.8590	(20)
	Ne II	30.4520	(20)
		30.5483	(20)
Argon	Ar Iα	11.6233	100
		11.8278	50
	Ar II	13.3019	30
		13.4794	15
Hydrogen	Lyman α	10.1986	100
	Lyman β	12.0872	10
	Lyman γ	12.7482	1

The relative intensities given are a rough guide for capillary discharge lamps. The intensities of the Ne II lines given in parentheses are from J. A. R. Samson, *Techniques of Vacuum Ultraviolet Spectroscopy*, John Wiley, New York (1967), and refer to a duoplasmatron light source.

2.2.2   LIGHT SOURCES WITH MONOCHROMATORS

Far ultraviolet grating monochromators can isolate individual lines from the sources already mentioned, or can be used with continuum or many-line light sources to provide light of variable energy. Useful laboratory lamps that give a wide energy range are the Hopfield continuum of helium[4] and the low-pressure capillary spark discharge in hydrogen[5]. Other useful lines can be isolated from the spectra of various ionized species in gas discharge lamps[6], but in all instances the intensities are much lower, even before monochromator losses, than those of the resonance lines usually used. In the energy region above 12 eV, no window material exists that will both transmit light and tolerate the working conditions of the lamps, so the vacuum monochromator and lamp must be differentially pumped.

The most promising light source for variable wavelength conditions is synchrotron radiation produced by electron accelerators or storage rings. Radiation is emitted as a consequence of the radial acceleration of electrons in a magnetic field, and is directed tangentially to the electron flight path. It has a continuous energy distribution in which the intensity increases with increasing energy up to a maximum at a few hundred volts, according to the energy of the electrons and the curvature of their path. Other characteristics are that the radiation is pulsed in pulses a few nanoseconds wide at a repetition rate of about 10 MHz, is almost fully polarized in the plane of the electron paths (80% or more) and has a very small angular divergence out of the plane. The intensity of present synchrotron sources seems to be just sufficient for photoelectron spectroscopy; it has already been used in photoionization and absorption studies of gases and of solids[7, 8].

2.2.3 VERY SOFT X-RAYS

Photoionization by X-radiation with energies of 1000 eV and higher is well established as a separate branch of photoelectron spectroscopy, which has been called ESCA (electron spectroscopy for chemical analysis) by Siegbahn et al.[9]. There are also some promising X-ray sources in the energy range 130–300 eV, which might help to bridge the gap between the ultraviolet and X-ray regions. These soft X-rays have been investigated by Krause[10], and the most useful one found is an yttrium line (M ξ) at 132.3 eV. This line unfortunately has an intrinsic width of about 0.6 eV, and its intensity in the sources so far developed has been very low.

2.3 ELECTRON ENERGY ANALYSERS

The electron energy analyser is the heart of any spectrometer, as it is here that the photoelectrons are separated according to their

Bessel box

kinetic energies. The aim of electron analyser design is to provide high resolution and high sensitivity simultaneously, requirements which naturally conflict. The theoretical resolution of any analyser can be increased by placing restrictions on the paths that electrons must follow in order to enter the analyser, but the gain in resolution achieved in this way results in a large decrease in intensity. Practical analysers must be designed with focusing properties such that they can accept as large a fraction as possible of the electrons produced in a properly matched source, while still giving the desired resolution.

The two major classes of analyser are those in which retarding fields are used to produce integral photoelectron spectra such as that in *Figure 1.1*, and those in which deflecting fields separate the electrons of different energies and give differential spectra.

2.3.1 RETARDING-FIELD ANALYSERS

Retarding-field analysers operate on the principle of permitting only those electrons which have energies higher than a retarding potential to reach the detector. Spectra are produced by recording the photoelectron current as the retarding potential is varied, and should contain a step in the current for each group of photoelectrons of discrete energy. The steps for low-energy electrons are super-imposed on the electron current of all electrons of higher energy, and as the statistical noise in the spectrum is proportional to the total electron current, the examination of low-energy electrons is sometimes difficult. However, these analysers have the important advantages of being almost equally sensitive to electrons of all energies, with perhaps a slight bias in favour of low-energy electrons, and of being able to accept electrons in a large solid angle or from extended sources. They are generally more sensitive than deflection analysers in terms of the collection efficiency for electrons, and this advantage is especially important at low energies, where differential analysers are usually least sensitive. *Figure 2.3* shows the designs of a number of different retarding-field analysers.

The cylindrical grid analyser (*Figure 2.3a*) is now of mainly historical interest, being the analyser used in the earliest work in molecular photoelectron spectroscopy[1]. It has the grave defect that electrons which are emitted at an angle to the electric field, that is, in a direction which is not perpendicular to the grids, are detected at too low an apparent energy, so that the steps produced in the spectrum are not sharp. In later retarding-field analysers, this defect is overcome in various ways. In the slotted grid analyser (*Figure 2.3b*), the cylindrical form is retained, allowing a long ionization region and a high photoelectron current. One grid is replaced by a pile of spaced discs[11], which transmit electrons only in a small range of angles about the normal to the photon beam. In the spherical retarding-field analyser[12] (*Figure 2.3c*), the grids and collector are spherical and the ionization region is point-like, so that

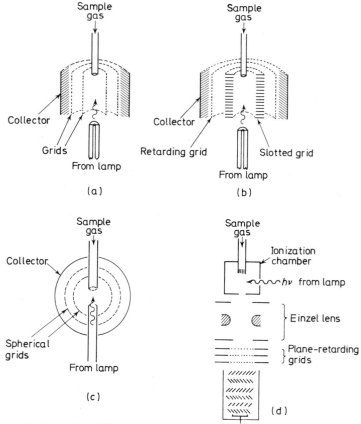

Figure 2.3. Retarding-field electron energy analysers: (a) cylindrical grid; (b) slotted grid; (c) spherical; (d) Einzel lens. All the analysers are rotationally symmetrical about a vertical axis

electrons ejected at all angles are collected and all of them automatically travel in the direction of the retarding field. The minimum number of electrodes in all of these retarding-field analysers is two, one to define the electrical potential at which electrons are formed and a second, which can also be the collector, to define a retarding potential. In most designs, a separate electrode for the retarding potential is provided, but some workers have reported that the highest resolution is attained with the minimum number of electrodes. The resolution depends on the uniformity of the fields and potentials and on the relative radii of the ionization volume and the surface defining the retarding potential. Electrons formed at points off the optic axis or centre can traverse the retarding field at an angle, and so appear to have a lower energy than they actually possess. The simplest approximation for the theoretical resolution is

$$\frac{\Delta V}{V} = \left(\frac{r}{R}\right)^2 \qquad (2.1)$$

where ΔV is the smallest resolvable energy difference at electron energy V, r is the radius of the ionization region and R that of the retarding electrode. Any lack of concentricity of the ionization source and retarding field contributes to r, so precise alignment is essential. More detailed theory on the resolution of spherical retarding-grid analysers has been given by Huchital and Rigden[13].

The most successful retarding-field analysers in terms of resolution are those based on the electron lens principle[14], as illustrated in *Figure 2d*. Ionization occurs at a point-like source, which may be defined by the intersection of the light beam with a beam of target gas, and the paths of electrons ejected into a certain solid angle defined by apertures are made parallel by an electron lens before the electrons impinge on a system of plane retarding grids. The electrons transmitted through the retarding field can conveniently be detected by using an electron multiplier. The resolution of this type of analyser depends on the excellence of the lens system and on the accuracy with which a retarding potential is defined by the grids; several grids are often provided in order to eliminate the effects of field penetration. A resolution of 5 meV has sometimes been attained but, as with all retarding-field analysers, it is extremely difficult to maintain high resolution in the presence of reactive target gases. Adsorption of gases on the grids causes local changes of contact potential, which are immediately and directly reflected in loss of resolution, because the electrons inevitably travel very close to the surfaces of the grid wires.

2.3.2   DEFLECTION ANALYSERS

In deflection analysers, electric or magnetic fields are used in order to make electrons of different energies follow different paths, and so to separate them. Magnetic analysers are not very suitable for low-energy electrons, because weak magnetic fields are needed and it is difficult to shield the analyser from stray fields present in the laboratory while applying a controlled field. Electrostatic analysers have come to predominate in photoelectron spectroscopy for this reason. There are many different types, each of which offers particular advantages and disadvantages, and most require a substantial mathematical treatment in order to describe their theoretical properties. Only the simplest, the 45 degree parallel-plate analyser will be described in detail, and then the important properties of the different analysers will be mentioned.

Figure 2.4 shows a parallel-plate analyser schematically. A uniform electric field, E, is established between two parallel plates

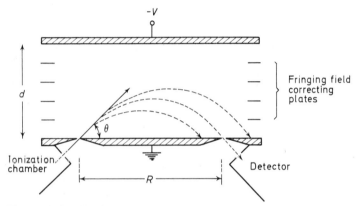

Figure 2.4. Parallel-plate electrostatic analyser in the 45 degree configuration

a distance d apart. Electrons of energy eV_0 and mass m_e enter the field at an angle θ by passing through a slit in the plate. The initial electron velocity, v_0, is given by

$$v_0 = (2\,eV_0/m_e)^{\frac{1}{2}} \tag{2.2}$$

and its components in the two directions x and y perpendicular and parallel to the field, respectively, are

$$v_{0x} = v_0 \cos\theta \tag{2.3}$$

$$v_{0y} = v_0 \sin\theta \tag{2.4}$$

The field is of such a polarity as to decelerate the electrons in the y direction, while it has no effect on their motion in the x direction. The equations of motion are therefore

$$\frac{dv_y}{dt} = -\frac{eE}{m_e} \tag{2.5}$$

$$v_y = v_0 \sin \theta - \frac{eE}{m_e} t \qquad (2.6)$$

$$y = v_0 t \sin \theta - \frac{eE}{2m_e} t^2 \qquad (2.7)$$

$$x = v_0 t \cos \theta \qquad (2.8)$$

The electrons follow parabolic trajectories, reaching a maximum height above the lower plate when $dy/dt = 0$, that is, when

$$t = \frac{mv_0}{eE} \sin \theta \qquad (2.9)$$

The second half of the trajectory is the mirror image of the first, so that the total time until the electrons return to the lower plate is just twice this value. The distance they have then travelled in the x direction, the range, is

$$R = 2v_0 \cos \theta \, \frac{mv_0}{eE} \sin \theta \qquad (2.10)$$

$$R = \frac{2eV_0}{eE} \sin 2\theta \qquad (2.11)$$

The condition for focusing is that the range should be independent of θ, that is, $\frac{dR}{d\theta} = 0$. This is fulfilled at $\theta = 45$ degrees, where $\sin 2\theta = 1$.

In a practical analyser, the range, R, is fixed by the positions of the inlet and entrance slits, and different electrons are brought into focus on the exit slit by varying the electric field. If a potential, V, is applied to the plates, the field is V/d, so that the operating condition of the analyser is

$$\frac{V}{V_0} = \frac{2d}{R} \qquad (2.12)$$

Hence, if the distance between the plates is exactly half the distance between the inlet and exit slits, the potential that must be applied to the plates to focus an electron is numerically equal to the energy of that electron in electronvolts.

A real analyser will have inlet and exit slits of finite widths S_1 and S_2, respectively, and the electrons will not all enter at exactly 45 degrees but in a range of angles near this value. The sensitivity of the analyser depends on the width of the slits and the range of angles that can be accepted, but the energy resolution also depends on these quantities. In practice, the resolution of photoelectron spectrometers is usually quoted in millielectronvolts as the width at half-height of the $^2P_{\frac{3}{2}}$ line of argon, near 5.5 eV electron energy. Theoretically, the energy width, ΔV, at an energy V is proportional to V for all deflection analysers, so a numerical resolution $V/\Delta V$ can be defined. For the 45 degree parallel-plate analyser, the resolution is given[15] by an equation of the form

$$\frac{\Delta V}{V} = \frac{S_1 + S_2}{R} + A\alpha^2 + \text{higher terms} \qquad (2.13)$$

where A is a constant and α is the fractional deviation of the electron entry angle from 45 degrees, when the angle is expressed as 45 $(1 + \alpha)$ degrees. Because the term in α vanishes, while that in α^2 is not zero, the 45 degree parallel-plate analyser is said to have first-order focusing. Equation 2.13 or a related equation applies to all deflection analysers, and in some instances the terms in α^2 also vanish, giving second-order focusing. The 45 degree analyser focuses electrons in only one plane (the plane of the paper in *Figure 2.4*) and this is called single focusing; other types of analyser focus in two planes simultaneously and are said to have double-focusing properties. An analyser with just first-order, single focusing can accept electrons only in a rather small solid angle while maintaining a given resolution. The actual angular deviation that is acceptable for a given resolution depends on the numerical value of the constant A, which for the 45 degree parallel-plate analyser is relatively large; it has weak first-order focusing.

A number of figures of merit for electron analysers have been proposed, such as the *luminosity*[16], *étendue*[17] or simply *transmission*. These all purport to be proportional to the signal strength that can be achieved for a given resolution and absolute source intensity, but are not, in fact, very helpful. Analysers are needed for different purposes, even within photoelectron spectroscopy, where each has its own special advantages. Furthermore, the energy range can always be scanned either by varying the field within the analyser, or by varying a pre-acceleration or deceleration of the electrons so as to bring them to a single fixed energy. Pre-acceleration can be a very simple procedure or involve subtle lenses with quadrupole or hexapole optics that match the characteristics of the electron source to the entrance aperture. These external arrangements or operating conditions have at least as strong an effect on the final performance of a spectrometer as the choice of analyser type. The forms of ionization region to which the different types of analyser can most advantageously be matched without lens systems are among their characteristics, which are discussed briefly in the following paragraphs.

45 degree parallel-plate analyser. The weak first-order single focusing of this analyser means that electrons can be accepted only within a small range of angles (± 3 degrees) for useful resolution[18]. The analyser can be matched to a line source of electrons, such as the pencil light beam from a capillary discharge lamp ionizing gas in a target chamber, but because of its poor collection efficiency it is not a good choice for normal photoelectron spectroscopy. In experiments where the angular deviation of the electrons must in any event be limited, as in studies of angular distributions, it is no worse than other analysers, and its advantages then come into play. These are its ease of construction and the fact that its inlet and exit slits are in an equipotential plane, which eliminates problems of fringing fields.

30 degree parallel-plate analyser. If the inlet and exit slits of a parallel-plate analyser are moved a calculated distance away from the positive plate and the electrons travel in a field-free space before

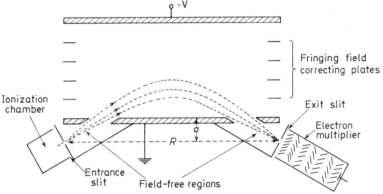

Figure 2.5. One configuration of the 30 degree parallel-plate electron energy analyser. When the exit and entrance slits are equidistant from the base plate of the analyser, the condition for second-order focusing is $R/a = 6\sqrt{3}$

entering the field at 30 degrees, an analyser with good second-order single focusing is obtained[19] (*Figure 2.5*). A further improvement can be made by placing the exit slit not at the position of second-order focus but at that of the minimum trace width* for a given angular deviation[20], and with this modification the 30 degree analyser

* The position of minimum trace width is that where the bundle of trajectories for electrons of a single energy but different initial angles is narrowest in a real analyser. It does not normally coincide with the position of the second-order focus.

is theoretically very attractive. It retains the advantages mentioned above for the 45 degree parallel-plate, but has the disadvantage that the large holes necessary in the first analyser plate must be covered with mesh for the sake of field uniformity. The mesh will cause a slight loss in transmission and introduces risks of contact potential effects.

Fountain analysers. Both the 45 and 30 degree parallel-plate analysers can be imagined in 2π geometry, with rotational symmetry about a point electron source, which they would then match. If the round plates of such an analyser are imagined to be lying in a horizontal plane, the electron trajectories would begin from a point below the centre of the lower plate, pass through a ring-like hole therein and spread outwards like water from a fountain before passing through a second ring-like hole to reach the collector. The 30 degree fountain analyser, modified for minimum trace width, has the best theoretical performance of any analyser proposed so far[20], but also has the serious disadvantage of an awkward-shaped electron collector that cannot act as a particle multiplier.

127 degree cylindrical analyser. The 127 degree cylindrical analyser[21] is the one most commonly used to date in photoelectron spectrometry. It has strong first-order focusing in one direction, and is suited to a line source of electrons. One problem is that the inlet

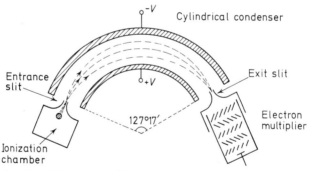

Figure 2.6. The 127 degree ($\pi/\sqrt{2}$) cylindrical analyser with electron trajectories, showing the use of curved entrance and exit slits in order to reduce the effects of fringing fields

and exit slits are not in an equipotential plane, and so precautions must be taken in order to prevent the presence of the slits from disturbing the field. Two solutions to this problem that have been used are to have specially shaped slit jaws, as shown in *Figure 2.6*, or to use the Herzog conditions[22], which reduce the total deflection angle from 127 to 119 degrees. A modification to the analyser that gives partial double focusing has been proposed[23] and the theory has recently been reviewed[24, 25].

Hemispherical analyser. The 180 degree hemispherical condenser analyser[26] (*Figure 2.7*) has first-order double focusing and matches a point source of electrons particularly well. The object and image points are again not in equipotential planes, so that a problem arises in the provision of slits. Simpson and Kuyatt[26] have provided a most elegant solution to this problem by having no real slits at the

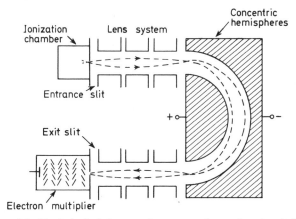

Figure 2.7. Hemispherical electrostatic energy analyser using virtual slits

analyser entrance or exit points, but lens systems which produce there the images of the real, physically distant slits. The theoretical performance of the hemispherical analyser with pre-acceleration of the electrons in the lens system is very good, and it is being used increasingly in photoelectron spectrometers.

Cylindrical mirror analyser. The system of coaxial cylinders as an energy analyser[27, 28] can have second-order focusing for angular deviations in the planes of the electron trajectories and perfect focusing, because of its geometry, for variations of the radial angle. This is theoretically the best analyser available for electrons from a point source, apart from the impractical 30 degree parallel-plate fountain. The focusing in the cylindrical mirror is good over a wide range of angles, and the 'magic' angle of 54 degrees 44 minutes, at which variations of angular distribution of electrons have no effect on intensities, can conveniently be used. The use of the cylindrical mirror analyser (*Figure 2.8*) as a photoelectron spectrometer is just now beginning[29]. Apart from its excellent focusing properties, it has the advantage that the entry and exit slits are at earth potential and field free. One disadvantage is that the object point, where the ionization region must be located, is rather deeply buried and inaccessible within the analyser construction. There is another version of the cylindrical mirror analyser, not illustrated here, in which the inlet and exit slits are at the surface of the inner cylinder, and both types can be used in photoelectron spectrometers.

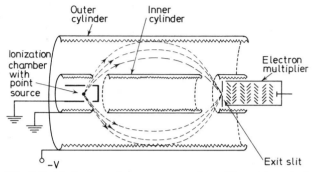

Figure 2.8. Cylindrical mirror electron energy analyser. The apertures in the inner cylinder must be interrupted for mechanical support, and covered with mesh for the sake of field uniformity. Fringing-field correcting plates must also be provided between the ends of the inner and outer cylinders as in parallel-plate analysers, but are omitted from the drawing for the sake of clarity

Other analysers. New electron analysers are being proposed at a high rate. Some old and new ones which may be of use in photo-electron spectrometry are the Wien filter[30], the tochoidal electron monochromator[31], the quadrupole energy filter[32] and the pill-box spectrometer[33]. The last example is an interesting departure from the usual practice of finding analytical solutions to the equations of motion, as was done in the design of all the other analysers. Instead, a mechanically simple design was chosen and the design parameters were worked out by numerical computer calculations of innumerable electron trajectories until a design with good focusing properties was found. Practical results are awaited.

2.4   ELECTRON DETECTORS AND RECORDING SYSTEMS

The electron currents encountered in photoelectron spectroscopy range from about 10 electrons per second (10^{-18} A) in differential spectrometers operated at the highest resolution, to 10^8 electrons per second (10^{-11} A) in the most efficient retarding-field spectro-meters. For currents higher than 10^{-14} A, simple electron collectors (Faraday cups) can be used with electrometer amplifiers. In an electrometer, the electron current develops a voltage across a very high resistor and this voltage is then amplified by special methods for display and recording. A good electrometer approaches closely the fundamental limit of sensitivity, which is set by the random thermal motions of electrons in the input resistor, called Johnson noise. This Johnson noise is additional to the statistical fluctuations that arise from collecting finite numbers of electrons, and limits the minimum currents that can be measured with any electrometer to about 10^{-16} A. For certain types of analyser, including most retarding-field analysers, the use of electron collectors and electrometers is necessary because of the awkward shape of the collectors needed. Electron collectors have the advantage of complete insensitivity to contamination by reactive gases and to bake-out at high temperatures, but this advantage is more than outweighed by their disadvantage of low sensitivity.

For the majority of photoelectron spectrometers, it is much more satisfactory to detect the photoelectrons singly by using an electron multiplier. When a single electron strikes the first electrode (dynode) of a multiplier with an energy of a few hundred electronvolts, it causes the emission of two or three secondary electrons, which are attracted to the second dynode. More secondary electrons are produced at the second dynode and after the same process has been repeated 10–20 times there are enough electrons to produce an easily measurable pulse in an external electrical circuit. Either the pulses can be counted digitally or the rate in pulses per second can be converted into analogue form in a ratemeter for display or recording.

The electron count-rate can be recorded on a pen recorder while the electron energy is scanned by variation of the analyser field or pre-accelerating potential. An X–Y recorder is normally used in which the vertical position of the pen is controlled by the count-rate and the horizontal position by the scanning voltage, so that the photoelectron spectrum is recorded directly. The speed of the scan and the electrical time constant of the recording circuit must be carefully chosen according to the resolution and the signal to statistical noise ratio desired, so that an appropriate number of electrons are detected at each point in the spectrum.

In a second system of recording, the spectrum is scanned rapidly many times, and the count of electrons detected at each energy is accumulated digitally in a multi-channel analyser. The contents of the multi-channel analyser can later be processed mathematically or read out on to an X–Y recorder. The advantages of this second method are that relative band intensities in the measured spectrum are not affected by slow changes of sensitivity, caused, for instance, by changes in the pressure of the sample gas, and that the problem of choosing the correct electrical time constant for a given scan speed is eliminated. The independence of the spectrum from sample pressure effects is important when only small amounts of sample are available, as it greatly increases the effective sensitivity. A disadvantage is that if variations of contact potentials change the effective energy scale during a measurement, the effect on the spectrum will be a reduction in the resolution. In the direct recording method, these very common energy scale drifts do not cause a loss in resolution but instead shift the peaks relative to one another and cause errors in the energy calibration.

2.5   THE OPERATION OF PHOTOELECTRON SPECTROMETERS

There are several precautions to be observed in measuring photo-electron spectra which require mention, although they are trivial in principle. The samples must be proved to be pure by techniques other than photoelectron spectroscopy, such as mass spectrometry, because the photoelectron spectrum itself may be a poor indicator of the presence of impurities. Furthermore, certain compounds can decompose within the target chamber, and many polar molecules may be partly dimeric under the flow conditions that exist there, even if they ought to be monomeric at equilibrium. A method of observing the mass spectrum and photoelectron spectrum concurrently has been proposed[34] as a general solution to these problems.

Even when the sample itself is pure, monomeric and undecomposed, the photoelectron spectrum does not necessarily represent the pure compound alone. A background spectrum is produced by scattered electrons and by foreign gases present in the source, the latter produced mainly by desorption from the walls. The usual causes are water, giving its characteristic peak at 12.6 eV ionization energy, or a previous sample being displaced by the new sample. Polar compounds are particularly strongly adsorbed, and the most peculiar spectra are sometimes seen when one compound displaces another from the surfaces of the inlet system and target chamber. It is good practice to allow a sample to flow through the spectrometer continuously for up to an hour before taking a spectrum, both to reduce these effects and to allow contact potentials to become stabilized. The background spectrum is definitely not independent of the presence of sample, so there is little point in measuring a background spectrum without the sample and subtracting it from the final observed spectrum.

The resolution of photoelectron spectrometers and the accuracy of energy and intensity measurements are partly a matter of spectrometer design, as described earlier in this chapter, but are also determined by the conditions under which the spectrometer operates. These topics are discussed in more detail in the following sections.

2.5.1 HIGH RESOLUTION

The theoretical resolution of any differential energy analyser can be made arbitrarily high by narrowing the slits and restricting the angular deviations of the electrons that enter the analyser. When this procedure is attempted, however, the experimental resolution is very seldom as good as the theoretical resolution, particularly at low electron energies. As the numerical resolution, $\Delta V/V$, should be constant, the instrumental peak width, ΔV, ought to be proportional to the electron energy, and this is usually found to be true for energies above about 5 eV. At lower energies, the peak width often tends to a constant value, and the numerical resolution becomes progressively lower as the electron energy decreases. The cause of this behaviour is probably not simple, but it is frequently attributed to local variations in surface potential and, hence, of the real potential difference between the target chamber and analyser. Surface conditions have a very strong effect on measured electron energies; variation of the pressure of even a rare gas can cause peak shifts of up to 0.5 V. The only way to prevent such effects from decreasing the resolution is to make all critical surfaces of the target chamber and analyser as electrically uniform as possible, which can be achieved either by covering all surfaces with a film of noble metal, usually gold, and then keeping them scrupulously clean, or by covering them with a layer of colloidal graphite. Apart from providing uniform electrical potentials, the graphited surfaces have the advantage of lower reflection coefficients than those of bare metals for low-energy electrons and thereby reduce the background signal of scattered electrons.

Two other risks to satisfactory resolution are the presence in the spectrometer of unwanted magnetic and electromagnetic fields. Low-energy electrons are remarkably sensitive to the presence of magnetic fields, and the longer the electron path length in the analyser the more serious the problem becomes. Both continuous and fluctuating magnetic fields must be eliminated, and this elimination can be attempted either by screening with high-permeability material (Mumetal) or by cancellation of the fields by using large and stable coil systems in which the currents are controlled by feedback from sensitive magnetometers. For high resolution work, all magnetic fields must be reduced to less than a few thousandths of the earth's field, and field gradients must also be made very small. Electromagnetic radiation, whether broadcast intentionally or unintentionally, can not only affect the electron trajectories directly but can also induce pick-up on any unscreened electrical leads, which act as antennas. In order to minimize such effects, the whole spectrometer and its electronic equipment must be earthed effectively, preferably at a single point, and all electrical connections to the spectrometer should be provided with a low impedance pathway to earth (decoupling).

The experimental or technical problems mentioned above are far more likely to limit resolution in practice than any fundamental physical effects. The few groups of workers who have achieved resolutions much better than 13 meV, which is offered by one commercial photoelectron spectrometer, have done so by paying painstaking attention to all these practical factors[35]. There is, however, one physical effect that limits resolution significantly, often called a Doppler effect. The target molecules before ionization have random thermal motions, and the components of their velocities in the direction of the spectrometer entrance slit are added to the electron velocities. The narrowest lines in a photoelectron spectrum have a width and shape characteristic of the molecular thermal velocity distribution, which can be derived from the kinetic theory of gases. With a differential electron energy analyser, which accepts electrons in a single direction only, the one-dimensional velocity distribution is relevant and the observable line width, ΔV, in millielectronvolts is given by

$$\Delta V = 0.75 \left(\frac{VT}{M}\right)^{\frac{1}{2}} \qquad (2.14)$$

where V is the electron energy in electronvolts, T the absolute temperature and M the mass in atomic mass units. The same equation is valid, with slightly different numerical factors[36], for analysers that accept electrons over wider angular ranges. For 10 eV electrons ejected from a molecule of mass 100 at room temperature, the energy spread, ΔV, is only 4 meV, which is well below normal analyser resolutions. It is for light molecules and high electron energies or gas temperatures that this Doppler effect becomes important, and the extreme case is hydrogen. Ionization of H_2 with He I light gives peaks 22 meV wide, and in ionization with He II light their width is 46 meV. The argon $^2P_{\frac{3}{2}}$ peak, which is often used as a standard test of resolution, has a Doppler width of

5 meV when ionization is by He I and 10 meV under He II ionization, all calculated at 300 °K.

It is clear that for work at very high resolution, or even at high
resolution when high temperatures or He II light are used, some
effort must be made to reduce the Doppler widths. Previous very
high resolution measurements have all been made by reducing the
electron energy as far as possible by using light of the longest usable
wavelength, because the $V^{\frac{1}{2}}$ term in equation 2.14 can thereby be
reduced by a large factor. Cooling the gas below room temperature is
impracticable for most substances, and can in any event have little
effect until temperatures below that of liquid nitrogen are reached.
A more generally attractive solution would be to provide a target
gas in the form of a jet or a true molecular beam, in which the
effective transverse kinetic temperatures used in equation 2.14 can
easily be reduced to below 10 °K. This method has not yet come into
use, but is expected to do so very shortly.

2.5.2   CALIBRATION AND ENERGY MEASUREMENTS

The energy scale of a photoelectron spectrum can be scanned either
by varying the potentials in the electron energy analyser or by
varying a potential difference between the target chamber and
analyser entrance slit. The second method has the advantages that
the analyser is set to transmit electrons of a fixed energy for which
its resolution and transmission can be optimized, and that the
resolution of the resulting spectrum is independent of electron
energy. Neither method, however, permits an absolute calibration
of the energy scale from instrumental parameters alone, and
calibration is always made by reference to the known ionization
potentials of some standard compounds. Contact potential effects
vary from compound to compound and shift the effective energy
scales as a function of pressure, so that it is essential to take the
spectrum of a mixture of sample and calibrant. When this is impossible, for instance, because the gases react chemically, spectra of
sample and calibrant have to be taken alternately, as rapidly as
possible, several times.

The gases used for calibration are chosen so as to give sharp peaks
in the spectrum at precisely known ionization potentials; the most
useful for the ionization energy range below 15 eV are argon, xenon
and methyl iodide. Some other gases and different ionizing radia-
tions from discharge lamps can be used in order to extend the range
to higher and lower ionization energies, and the most useful are
given in *Table 2.3* (taken mainly from the work of Lloyd[37]). Once

the spectrum of sample with calibrant has been obtained, the usual method of fixing the energy scale and determining ionization potentials is to make a linear interpolation between calibrant lines, based on the potential which is scanned experimentally. Lloyd[37] has shown, however, that in his spectrometer (127 degree analyser), exact linearity between analyser potential and electron energy cannot be relied upon in the region of electron energy below 5 eV. Deviations from exact linearity are probably more common than is often assumed, and for this reason the calibrant lines should always be as near as possible in energy to an ionization peak whose position is to be determined.

Table 2.3 USEFUL CALIBRATION LINES

Substance	Ionic state	Ionizing line	Electron energy, eV	Apparent* ionization potential, eV
Ne	$^2P_{3/2}$	He IIα	19.249 4	1.968 1
Ne	$^2P_{1/2}$	He IIα	19.346 3	2.065 0
He	^2S	He IIα	16.226 8	4.990 7
MeI	$^2E_{1/2}$	He Iβ	12.922	8.296
MeI	$^2E_{3/2}$	He Iα	11.680	9.538
MeI	$^2E_{1/2}$	He Iα	11.053	10.165
Xe	$^2P_{3/2}$	He Iα	9.088	12.130
Xe	$^2P_{1/2}$	He Iα	7.782	13.436
Kr	$^2P_{3/2}$	He Iα	7.219	13.999
Kr	$^2P_{1/2}$	He Iα	6.553	14.665
Hg	$^2D_{5/2}$	He Iα	6.378	14.840
Ar	$^2P_{3/2}$	He Iα	5.459	15.759
Ar	$^2P_{1/2}$	He Iα	5.281	15.937
Hg	$^2D_{3/2}$	He Iα	4.514	16.704
N_2	$^2\Sigma_u^+$	He Iα	2.467	18.751
Ne	$^2P_{3/2}$	He Iβ	1.522 3	19.695 2
Ne	$^2P_{1/2}$	He Iβ	1.425 3	19.792 2
CH_3I	$^2E_{3/2}$	H Lyman α	0.661	20.557

* As if the ionization were by He Iα radiation in all instances.

2.5.3 INTENSITY MEASUREMENTS

In differential photoelectron spectra, the relative heights of two bands that have intrinsically different widths, such as the first two bands in the spectrum of methanol (*Figure 1.11*), vary with the resolution of the analyser. As the resolution improves, the height of sharp peaks apparently increases while the height of broad bands is reduced. For a truly narrow peak, all the electrons can be transmitted through the analyser at a single setting, because their energies are all within the energy width, ΔV, which defines the resolution; if the true energies cover a range broader than ΔV, the fraction of them that can pass through the analyser is proportional to ΔV itself. Peak heights are therefore not an appropriate measure of intensity, and it can easily be shown that areas rather than heights should always be used. With photoelectron spectra from retarding-field analysers, this problem does not arise as the measured step heights in integral spectra correspond exactly to areas in differential spectra.

A more serious difficulty arises from variations of analyser sensitivity with energy, generally called energy discriminations, which distort the spectra and must be allowed for. All photoelectron spectrometers have some energy discriminations; simple deflection instruments are more sensitive to high-energy than low-energy electrons and retarding-field instruments often favour electrons of the lowest energy. It is extremely difficult to measure the energy discriminations experimentally, because there are so few photo-electron spectra or other sources of electrons for which precise intensity information is available. Instead, it is usual to accept the theoretical variations of sensitivity with energy for each type of analyser, and to use it to correct measured spectra. For deflection analysers without pre-acceleration of the electrons, the resolution, ΔV, and thus the energy bandwidth within which electrons are transmitted, is proportional to the electron energy, V. The areas of both sharp peaks and broad bands are proportional to ΔV, so that their measured intensities can be corrected by dividing them by the electron energy. Retarding-field analysers and deflection analysers in which pre-acceleration is used for scanning are assumed to have no energy discrimination, but in both instances with the constraint that at the low electron energy end of the scale the measured intensities are probably inaccurate. Deflection analysers with pre-acceleration are free from discrimination only if the solid angle within which electrons from the ionization region are collected is geometrically fixed before pre-acceleration; otherwise, this solid angle, and therefore the intensity, depend on the accelerating field.

The above discussion of angles is related to the final and most troublesome problem in correlating measured photoelectron spectra with the true photoelectron energy distribution produced by ionization. The electrons are not ejected in equal numbers in all directions and their angular distributions vary from one band to another in the photoelectron spectrum. In order to correct for this effect, it is necessary either to know or measure the angular distribution or else to collect all the electrons, as in the spherical retarding-field analyser. The theory of the angular distributions allows one other method, which is to make measurements at an angle of 54 degrees 44 minutes from the photon beam (for unpolarized light), because at this 'magic' angle variations in the angular distribution have no effect on the intensities. Angular distributions and their measurement are discussed more fully in the next chapter; here it may be noted that their effect on intensities is generally small. For a normal photoelectron spectrometer in which electrons normal to the photon beam are examined, the maximum possible variation in intensity due to angular effects alone is by a factor of two, and in practice the variations are normally much less.

2.6 SAMPLE AND VACUUM REQUIREMENTS

Because photoionization cross-sections are relatively small (10–150×10^{-18} cm^2), even the high intensity of the resonance lamps is not sufficient to make photoelectron spectrometry a very sensitive technique. The pressure in the target chamber needed to measure a useful spectrum is 10^{-3}–10^{-1} torr, and the amount of sample required is correspondingly of the order of milligrams rather than micrograms. This limits the substances that can be examined without heated inlet systems to those with rather high vapour pressures at room temperature, and for compounds less volatile than, say, naphthalene, special techniques are required. The following three methods have been used to study less volatile materials.

(1) The whole spectrometer, including the inlet system, can be heated, but temperatures higher than about 100 °C are not possible if the construction includes rubber gaskets or if electron multipliers are used, as their background count-rate increases sharply with temperature. If a multiplier is not used, the analyser can be heated up to about 350 °C and solid samples can be inserted into it directly on a probe. It is difficult to find materials that retain sufficiently good insulating properties at such a high temperature for the electron collector to be isolated without degrading the performance of the electrometer, and pure aluminium oxide seems to be the best choice. Another problem is that many solids are thermally ionized even at 300 °C, and can give rise to a strong background signal. Despite these problems, a number of involatile

compounds have been examined in this way, including large aromatic hydrocarbons[38] and lead, tin and cadmium halides[39].

(2) The inlet system and the target chamber alone can be heated, and vapour allowed to escape into the body of the cold spectrometer by passing it through the exit slit of the target chamber. The pressure in the target chamber is 10–100 times higher than that in the analyser because of differential pumping, so that condensation on the spectrometer surfaces need not be a problem. When very involatile substances are examined in this way, however, condensation is bound to occur and the analyser must be cleaned periodically. Care must be taken that the multiplier does not become contaminated, because the dynodes are very sensitive and multipliers are expensive. Many involatile compounds have been examined by this method following the commercial availability of a heated inlet system that operates on this principle. If the vapours to be studied are brought into the spectrometer from the exterior, much can be gained by making the pipes that carry them as short, wide and direct as possible[40]. The sample is continually being pumped away through the exit slit of the ionization chamber, and the throttling effect of a long connecting tube can reduce the pressure in the target chamber much below the vapour pressure of the sample at a given temperature.

(3) The target gas can be supplied in the form of a jet or beam from an oven, crossed with the photon beam in a well pumped target chamber and later condensed on a designated, preferably cooled, surface. This method offers the widest range of temperature and volatility and is unquestionably the best way of examining high-temperature vapours. It has the advantage that the ordering of molecular motions in the jet or beam can reduce the Doppler broadening of photoelectron peaks, which might otherwise be severe at high temperatures. The beam method has already been used to measure the photoelectron spectra of thallium and lead halides[41], and there is no reason why even such involatile salts as sodium chloride should not be examined by its use[42].

Under normal working conditions, when involatile substances are not involved, the pressure in the vacuum vessel of a photoelectron spectrometer may easily reach 3×10^{-4} torr. The normal working conditions often approach closely the point at which the electron signal is no longer proportional to the pressure and where peaks are broadened because electron scattering processes become significant. This high-pressure region must, of course, be avoided in accurate work, but the range of pressure involved is sufficient to show that the most stringent conditions of high-vacuum technology need not be used in photoelectron spectroscopy. Oil diffusion pumps without cold traps are normally satisfactory, and brass, aluminium and rubber gaskets and other poor high-vacuum materials can be used without adverse effects. Some physicists pursue a different policy and use no graphite or even gold coatings in their photo-electron spectrometers, but instead maintain very clean metal surfaces, often of stainless steel. This approach requires the use of much better vacuum conditions, cold traps and perhaps mercury diffusion pumps, and the very careful examination of every material to be used within the vacuum system[43]. This policy is essential if surface effects are to be examined, but is not generally necessary when only gaseous samples are studied. The ultra-high vacuum conditions needed for the preparation and study of atomically clean surfaces are outside the scope of this book, but are briefly mentioned again in Chapter 8.

REFERENCES

1. AL-JOBOURY, M. I. and TURNER, D. W., *J. chem. Soc.*, 5141 (1963)
2. SAMSON, J. A. R., *Rev. scient. Instrum.*, **40**, 1174 (1969)
3. POTTS, A. W., WILLIAMS, T. A. and PRICE, W. C., *Discuss. Faraday Soc.*, **54**, 104 (1973)
4. HUFFMAN, R. E., TANAKA, Y. and LARRABEE, J. C., *Bull. Amer. Phys. Soc. II*, **7**, 457 (1962)
5. WEISSLER, G. L., *Handbuch der Physik*, Vol. XXI, Springer Verlag, Berlin (1956)
6. SAMSON, J. A. R., *Chem. Phys. Lett.*, **12**, 625 (1972)
7. MADDEN, R. P. and CODLING, K., *Phys. Rev. Lett.*, **10**, 516 (1963)
8. KOCH, E. E. and SKIBOWSKI, M., *Chem. Phys. Lett.*, **9**, 429 (1971)
9. SIEGBAHN, K., NORDLING, C., FAHLMAN, A., NORDBERG, R., HAMRINN, K., HEDMAN, J., KOHANSSON, G., BERGMARK, T., KARLSSON, S. E., LINDGREN, I. and LINDBERG, B., *Nova Acta Regiae Soc. Sci. Upsal., Ser. IV*, **20** (1967)
10. KRAUSE, M. O., *Chem. Phys. Lett.*, **10**, 65 (1971)
11. PRICE, W. C., in Hepple, P. (Editor) *Molecular Spectroscopy*, Institute of Petroleum, London, 221 (1968)
12. FROST, D. C., MCDOWELL, C. A. and VROOM, D. A., *Proc. R. Soc., Lond.*, **A296**, 566 (1967)
13. HUCHITAL, D. A. and RIGDEN, J. D., in Shirley, D. A. (Editor) *Electron Spectroscopy*, North Holland, Amsterdam, 79 (1972)
14. SPOHR, R. and VON PUTTKAMER, E., *Z. Naturforsch.*, **22a**, 409 (1967)
15. RUDD, M. E., in Sevier, K. D. (Editor) *Low Energy Electron Spectroscopy*, John Wiley, New York (1972)
16. AKSELA, S., KARRAS, M., PESSA, M. and SUONINEN, E., *Rev. scient. Instrum.*, **41**, 351 (1970)
17. HEDDLE, D. W. O., *J. Physics, E, scient. Instrum.*, **4**, 589 (1971)
18. HARROWER, G. H., *Rev. scient. Instrum.*, **26**, 850 (1955)
19. GREEN, T. S. and PROCA, G. A., *Rev. scient. Instrum.*, **41**, 1409 (1970)
20. SCHMITZ, W. and MELHORN, W., *J. Physics, E, scient. Instrum.*, **5**, 64 (1972)

21. HUGHES, A. L. and ROJANSKY, V., *Phys. Rev.*, **34**, 284 (1929)
22. HERZOG, R., *Z. Physik*, **97**, 586 (1935)
23. LEVENTHAL, J. J. and NORTH, G. R., *Rev. scient. Instrum.*, **42**, 120 (1971)
24. JOHNSTONE, A. D., *Rev. scient. Instrum.*, **43**, 1030 (1972)
25. ARNOW, M. and JONES, D. R., *Rev. scient. Instrum.*, **43**, 72 (1972)
26. KUYATT, C. E. and SIMPSON, J. A., *Rev. scient. Instrum.*, **38**, 103 (1967)
27. ZASHKVARA, V. V., KORUNSKII, M. I. and KOSMACHEV, O., *Sov. Phys.-Tech. Phys.* [English translation], **11**, 96 (1966)
28. AKSELA, S., *Rev. scient. Instrum.*, **42**, 810 (1971)
29. BERKOWITZ, J., *J. chem. Phys.*, **56**, 2766 (1972)
30. ANDERSON, W. H. J. and LEPOOLE, J. B., *J. Physics, E, scient. Instrum.*, **3**, 121 (1970)
31. STAMATOVIC, A. and SCHULZ, G. J., *Rev. scient. Instrum.*, **41**, 423 (1970)
32. SCHMIDT, C., *J. Physics, E, scient. Instrum.*, **5**, 1063 (1972)
33. ALLEN, J. D., WOLFE, J. P. and SCHWEITZER, G. K., *J. Mass Spectrom. Ion Physics*, **8**, 81 (1972)
34. AMES, D. L., MAIER, J. P., WATT, F. and TURNER, D. W., *Discuss. Faraday Soc.*, **54**, 277 (1973)
35. ÅSBRINK, L. and RABALAIS, J. W., *Chem. Phys. Lett.*, **12**, 182 (1971)
36. TURNER, D. W., *Phil. Trans. R. Soc., Lond.*, **A268**, 7 (1970)
37. LLOYD, D. R., *J. Physics, E, scient. Instrum.*, **3**, 629 (1970)
38. DEWAR, M. J. S. and GOODMAN, W., cited by Worley, S. D., *Chem. Revs.*, **71**, 295 (1971)
39. COCKSEY, B. J., DANBY, C. J. and ELAND, J. H. D., to be published
40. EVANS, S., ORCHARD, A. F. and TURNER, D. W., *J. Mass Spectrom. Ion Physics*, **7**, 261 (1971)
41. BERKOWITZ, J., in Shirley, D. A. (Editor) *Electron Spectroscopy*, North Holland, Amsterdam, 391 (1972)
42. EVANS, S., *Discuss. Faraday Soc.*, **54**, 143 (1973)
43. WHEELER, W. R., *Physics Today*, **25**, No. 8, 52 (1972)

3 Ionization

3.1 INTRODUCTION

A more detailed examination of photoelectron spectroscopy must begin with a study of the process of ionization itself. Ionization can be brought about by photons either in the single-step process of direct photoionization or in a two-step process, autoionization. Although autoionization does not often manifest itself by affecting photoelectron spectra directly, it does sometimes do so. It interferes very seriously with the study of ionization by almost all other techniques and so contributes indirectly to the importance of photoelectron spectroscopy. A photoelectron spectroscopist should nevertheless have some familiarity with these other methods of ion physics and chemistry, and some of them are introduced briefly in this chapter.

3.2 PHOTOIONIZATION

In the study of electronic absorption and emission spectra of atoms and molecules, much importance is attached to the selection rules that indicate which transitions are allowed and which are forbidden in terms of various quantum numbers. Although photoionization is an electronic transition and is an interaction with electric dipole radiation, there is no need to take account of a large number of selection rules that describe it. For all common photoionization processes, one rule suffices, namely that they are *one-electron transitions*. Normal photoionization is the removal of a single electron from the neutral species without changing the quantum number of any other electrons, and this is allowed whatever orbital the electron is in. For an atom or linear molecule, if the angular momentum of the original electron is l then, considering the change between the neutral molecule and the ion, the angular momentum change, ΔL, is given by

$$\Delta L = 0, \pm 1 \ldots \pm l \qquad (3.1)$$

As one electron with spin $\pm\frac{1}{2}$ is removed, the rule for the change in multiplicity is automatically

$$\Delta S = \pm 1 \qquad (3.2)$$

Equations 3.1 and 3.2 are not rules that somehow restrict the possibilities of photoionization, but are simply a description of the difference between the molecule and the ion formed from it by removal of one electron without disturbing the others. This catholicity of photoionization is due to the fact that the final state of the system is the molecular ion plus a free electron. The change in the whole system between molecule plus photon and ion plus electron is restricted by the usual dipole selection rules, but the free electron can leave the ion carrying whatever angular momentum is needed to satisfy these rules. The motion of the free electron is described by a wave-function, and it can be an s, p, d or f wave, carrying zero, one, two or three units of angular momentum, respectively. In order to fulfil the selection rules, the electron needs an angular momentum of $l \pm 1$ and when two different angular momenta of the electron are possible by this rule both may be utilized, giving an outgoing wave of mixed character. If the original electron is in a molecule with low symmetry where angular momentum is not defined, the outgoing electron wave-function is again a mixture of s, p, d or f waves, and equation 3.1 can be replaced by

$$\Delta \Gamma = \times \Gamma_j \qquad (3.3)$$

This equation states that the change in symmetry species between molecule and ion is obtained by taking the direct product of the species of the molecule with that of the electron to be ionized. For rotational changes that accompany ionization, the most intense transitions are those with $\Delta J = 0, \pm 1$, but this is not a strict selection rule and higher ΔJ values are also possible (see Section 3.4.3).

The only forbidden processes of importance in photoelectron spectroscopy are two-electron processes, such as the ionization of one electron and simultaneous excitation of another, or the ejection of two electrons from a molecule by a single photon. The forbidden nature of these processes arises in a different way from the usual selection rules, which are based on symmetry; it is derived from the fact that the electronic structure of most molecules or atoms is well described by the familiar one-electron orbital model. It can be proved that if the motions of electrons are independent of one another, which is the fundamental assumption of the molecular orbital model, a transition induced by radiation that changes the quantum numbers of one electron must leave the quantum numbers of all other electrons unchanged[1]. Two-electron transitions occur because the motions of electrons are not completely independent but are correlated, and the intensity of two-electron transitions is a measure of this electron correlation. An example of a two-electron transition in photoelectron spectroscopy is the very weak band found in the spectrum of atomic mercury for the process[2]

$$Hg\,(5d^{10}\,6s^2\ldots{}^1S)+h\nu \rightarrow Hg^+\,(5d^{10}\,6p\ldots{}^2P)+e \qquad (3.4)$$

Another two-electron excitation occurs in the photoelectron spectrum of hydrogen ionized by 247 Å light[3]:

$$H_2\,(1s\sigma^2\ldots{}^1\Sigma_g^+)+h\nu \rightarrow H_2^+\,(2s\sigma_u\ldots{}^2\Sigma_u^+)+e \qquad (3.5)$$

The ejection of two electrons by a single photon, called photo-double ionization, has been studied in the rare gases[4], and the theory has been developed[5]. The excess energy of the photon above the double ionization threshold can be distributed between the two electrons that leave the molecular ion, and no sharp peak is observed in the photoelectron spectrum. Instead, the electrons have a continuous distribution of energies with two maxima, one at zero energy and the other at the position of the excess energy available. The thresholds for photo-double ionization of molecules are almost all higher than 21.22 eV, so these continua do not interfere with normal photoelectron spectra.

Because all one-electron photoionizations are allowed, the partial photoionization cross-sections, or the probabilities of ionization to particular states of ions, are all of the same order of magnitude. There are variations within a factor of two or three between one doubly occupied orbital and another, as already mentioned in Chapter 1, and there are also variations with the wavelength of the ionizing radiation. Photoionization is not a resonance process, and once the threshold for a given ionization has been passed, light of all shorter wavelengths can cause the same process, the excess energy appearing as kinetic energy of the electrons. The cross-section is usually highest near the threshold, after which it decreases with increasing photon energy. When the photon energy is very high compared with the threshold energy, the cross-section should be proportional to v^{-3}, but this does not apply in the energy region used in ultraviolet photoelectron spectroscopy. Experimental measurements of the partial ionization cross-sections as a function of energy are needed, but involve severe practical difficulties and have been made only for the rare gases,[6] atomic mercury[7] and a few small molecules[8-10]. All these studies show that the partial ionization cross-sections for direct photoionization vary little between threshold energies of about 10 eV and the normal photon energy of 21.22 eV, and the variation that exists is not sufficient to invalidate the limiting rules for relative band intensities. This is also confirmed by the observation that there is seldom a large change in the spectrum of a molecule when the ionizing radiation is changed from He I to He II. Molecules that contain both light and heavy atoms provide an exception to this, as the cross-sections for photoionization from heavy atom orbitals decrease much more rapidly than those of light atom orbitals between the He I and He II wavelengths.

Before the advent of photoelectron spectroscopy, photoionization was studied by measuring the yields of ions and electrons, the photocurrents, produced by the passage of ultraviolet light of different wavelengths through gases. If direct photoionization were the only process of importance, curves of the photocurrent as a function of the light energy would contain the same information as photoelectron spectra, and would indeed be almost identical in form with integral photoelectron spectra. The reason why this is not so, in fact, is the very widespread occurrence of autoionization processes, which are considered in the next section.

3.3 AUTOIONIZATION

In contrast to direct ionization into the continuum, which forms the basis of photoelectron spectroscopy, autoionization is an indirect process. Molecules are first produced in a neutral excited state with energy above an ionization limit, and then spontaneously emit electrons:

$$M + h\nu \rightarrow M^* \rightarrow M^+ + e \qquad (3.6)$$

The first step is a resonance process; it can be brought about only by light of the correct energy, and the nature of the accessible excited states is governed by the normal optical selection rules. Autoionization proper is the second step, and as no radiation is involved it follows the monopole selection rule that the symmetry species of the autoionizing state and the final state of the ion plus free electron must be identical. Because the energy available in the process of equation 3.6 is still the photon energy, the kinetic energy of the electrons ejected in the second step is an energy which the photo-electrons could have as a result of direct photoionization at the same wavelength. The intensities of the different bands and vibrational lines may, however, be very different.

Autoionizing states are detected in experiments in which the photon energy is varied, and they appear as peaks or sometimes as

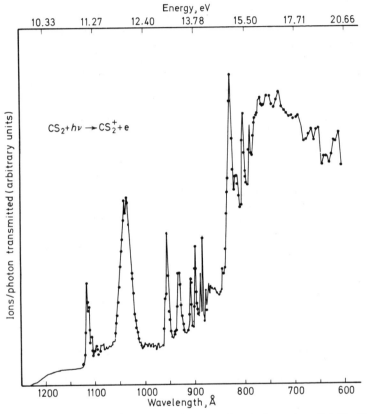

Figure 3.1. Photoionization cross-section curve for the formation of stable CS_2^+ ions from carbon disulphide. (From Dibeler, V. H. and Walker, J. A., *J. opt. Soc. Amer.*, **57**, 1007 (1967), by courtesy of the Optical Society of America)

dips in graphs of photoionization current against wavelength. The features due to autoionization are often so intense and numerous that they completely hide the structure that represents direct photoionization, as they do, for instance, in the photoionization cross-section curve shown in *Figure 3.1*. Autoionization may be more important for small than for large molecules, but nevertheless no case is yet known in which the photoionization cross-section curve is so free from autoionization structure that it is equivalent to an integral photoelectron spectrum.

If the frequency of the ionizing radiation used in photoelectron spectroscopy matches an autoionizing state, the photoelectron spectrum may be distorted. One of the advantages of photoelectron spectroscopy is that this happens infrequently, whereas in techniques in which the energy of the ionizing particles, photons or electrons is varied, all autoionizing states are encountered. The autoionizing states are generally Rydberg states of the neutral molecule belonging to series that converge on some higher limit, and they are usually found in the region of about 2 eV below the limit on which they converge. Because most valence electron ionization potentials are less than 21 eV, the autoionizing states become much less common at the wavelength of He I, and are rarely encountered at that of He II. The situation is very different with the neon resonance lines at 744 and 736 Å and with the argon lines at 1067 and 1048 Å, where autoionizing states are found in many molecules. There is always a high chance that autoionization will complicate a spectrum obtained with light of these longer wavelengths.

There are two distinct types of autoionization, according to whether the electron that is eventually ejected is the same electron as that excited in the resonance process, or whether it comes from another orbital of the neutral molecule. In the first instance, called vibrational autoionization, the autoionizing state must have an electron in a Rydberg orbital with a high principal quantum number and also enough vibrational excitation in the molecular ion core to bring the total energy above the limit. Autoionization follows by the interaction of vibrational and electronic motions in which the vibrational energy of the core is converted into electronic energy, and the Rydberg electron is ejected. This type of autoionization has been studied theoretically[11], and also experimentally[12], for hydrogen. The smallest possible change of vibrational quantum number in the autoionization step is most probable, and so the ejected electrons are of almost zero energy. The prevalence of this process in photoelectron spectroscopy is difficult to judge, because normal photoelectron spectrometers cannot be relied upon at electron kinetic energies near zero.

In the second type of autoionization, sometimes called electronic autoionization, the molecular ion is formed in a different electronic state from that of the core of the autoionizing level. There is no restriction to the ejection of an electron with low kinetic energy, and any final state of the ion with a lower energy than that of the autoionizing state can be produced. This form of autoionization can

produce two types of distortion in the photoelectron spectrum, as follows:
(1) Changes in the vibrational structure of electronic bands. Both the formation of the autoionizing state and its ionization are separately governed by the Franck–Condon principle, so the shape of the vibrational structure in the photoelectron spectrum depends on the equilibrium internuclear distances in the autoionizing state, as well as in the ground-state molecule and the final ionic state. A possible situation is illustrated in *Figure 3.2*, which shows how vibrationally excited states of a

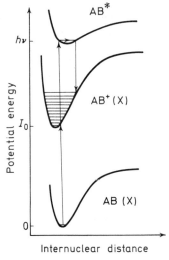

Figure 3.2. Potential energy diagram to illustrate how electronic autoionization can populate highly excited vibrational levels of an ionic state

molecular ion that cannot be reached in direct ionization can be populated by the indirect process. The best example of this is the dramatic change in the photoelectron spectrum of oxygen between He I and Ne I radiation, shown by *Figure 3.3*. Several other examples are known from the photoelectron spectra of small molecules excited with neon or argon resonance radiation, and nitrogen[12], oxygen[13] and nitric oxide[14] have been examined in more detail at chosen autoionizing resonances. The theory of the changes in vibrational intensities has also been discussed[15]. It is possible that spectra taken at 584 Å contain more examples of this type of distortion than is usually assumed, because it is impossible to detect unless a spectrum taken at a different wavelength is available for comparison.

(2) Changes in relative band intensities. If the autoionizing state has a higher energy than those of several ionic states, the branching ratios to the different states may not be the same as in direct ionization. The selection rules do not help in the estimation of the branching ratios, because any state reached in direct ionization by an allowed one-electron transition can also be reached by autoionization from neutral states populated by an allowed absorption. Nor can ionic states be

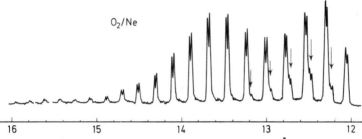

O_2/Ne

16   15   14   13   12

Ionization energy, eV, for ionization by the 736 Å line

Figure 3.3. $^2\Pi_g$ band in the photoelectron spectrum of oxygen excited by neon resonance radiation, to be compared with *Figure 1.6*. The extensive vibrational structure is due to autoionization caused by the strong neon line at 736 Å. The weaker line at 744 Å gives only normal direct ionization, producing the peaks marked with arrows. The doublet structure of the peaks is due to spin–orbit coupling. (By courtesy of Professor W. C. Price)

produced by autoionization that are not reached in direct ionization unless the original absorption is a two-electron excitation. No theoretical predictions of branching ratios in electronic autoionization have yet been made, although a theoretical formalism exists; experimental results are also few. Blake and Carver[7, 8, 10] have measured partial photo-ionization cross-sections as a function of wavelength for a number of atoms and small molecules, and the presence of autoionizations with different branching ratios is apparent from their data. Unfortunately, their photon energy resolution (about 10 Å) is not sufficient for the individual autoionizing lines to be separated, except for mercury[7]. In the curves for mercury (*Figure 3.4*) the autoionization processes stand out as sharp peaks on the almost flat continua for direct photo-ionization. All the peaks in the partial cross-section for the formation of Hg^+ ($^2D_{\frac{5}{2}}$) represent autoionizing levels that can also decay to Hg^+ ($^2S_{\frac{1}{2}}$); the relative intensities show that they do not decay to both $^2D_{\frac{5}{2}}$ and $^2S_{\frac{1}{2}}$ impartially. The same autoionizing lines are present in *both* partial ionization cross-sections and the variation in their relative

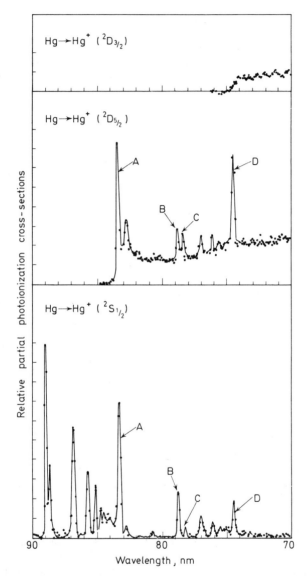

Figure 3.4. Partial photoionization cross-sections as a function of wavelength for the production of the three states $^2S_{\frac{1}{2}}$, $^2D_{\frac{5}{2}}$ and $^2D_{\frac{3}{2}}$ of Hg$^+$. Some corresponding autoionization peaks in the $^2S_{\frac{1}{2}}$ curves are marked A, B, C and D, and show variations in the branching ratios for autoionization. (From Blake[7], by courtesy of the Council of the Royal Society)

intensities proves that branching ratios from different auto-ionizing levels are different, and therefore they cannot all be the same as the branching ratio in direct photoionization. The direct photoionization process that forms Hg^+ ($^2S_{\frac{1}{2}}$) has such a low cross-section that it was not detected at all in the experiments that led to *Figure 3.4*.

When differences in relative band intensities are found between photoelectron spectra taken at two photon energies, it is sometimes possible to conclude that different branching ratios in electronic autoionization are their cause. A striking change occurs in the photoelectron spectrum of sulphur hexafluoride between excitation by He I and He II light, and is illustrated in *Figure 3.5*. The changes

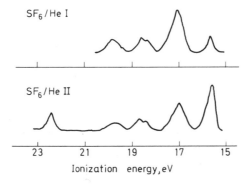

Figure 3.5. Photoelectron spectra of sulphur hexafluoride excited by He I and He II light. (From Price, W. C., Potts, A. W. and Streets, D. G., in Shirley, D. A. (Editor) *Electron Spectroscopy*, North Holland, Amsterdam (1972))

in relative intensity are so strong, and the intensities at 584 Å deviate so much from the predictions of the limiting rules for intensities (Chapter 1, Section 1.3), that in this instance electronic autoionization at 584 Å seems a likely cause. There is another ionization limit at 22.5 eV, and Rydberg series that converge on this limit might well include a line at 584 Å. It is an open question how many other photoelectron spectra taken at 584 Å are similarly distorted, but probably the answer will be that very few are. Changes in experimental relative band intensities with wavelength are also caused by variations in the direct photoionization cross-sections and in the angular distributions of the photoelectrons, and such changes are more common than those caused by electronic autoionization.

In addition to normal autoionization, a more complex auto-ionization process may exist that involves fluorescence emission:

$$M + h\nu \rightarrow M^{**} \rightarrow M^* + h\nu_{fluor.} \tag{3.7}$$

$$M^* \rightarrow M^+ + e \tag{3.8}$$

Here the excited state of the neutral species that are initially formed in absorption emits fluorescence radiation, but still retains enough excitation energy to autoionize. This is the first process to be considered in which the normal energy balance of equation 1.4 does not hold; instead,

$$KE = h\nu - h\nu_{fluor.} - I_j - E^*_{vib., rot.} \tag{3.9}$$

The measured electron energies are no longer directly related to the energies of the ionic states produced, because of the energy lost in fluorescence. This mechanism can explain the appearance of bands in photoelectron spectra taken at certain wavelengths that do not otherwise correlate with any known ionic state, and it was proposed for this purpose by Blake and Carver[10]. In a normal band in a photoelectron spectrum, the electron energy must increase continuously as the energy of the light is increased, the difference $h\nu - KE$ remaining constant. In the proposed fluorescence auto-ionization, the electron energy may behave in a complicated manner, perhaps remaining virtually constant over a range of photon energies as different vibrational levels in the fluorescent autoionizing state are reached, or as a single autoionizing state is populated by fluorescence emission from several higher states. The process of fluorescence autoionization has been proposed as an explanation of bands that show this behaviour in the photoelectron spectra of oxygen[10] and ammonia[8], but its existence cannot yet be taken as established.

3.4 ANGULAR DISTRIBUTIONS OF PHOTOELECTRONS

In photoionization, the photoelectrons are not emitted equally in all directions, nor are their angular distributions the same when different ionic states are produced. Band areas in experimental photoelectron spectra therefore depend on the angle with respect to the light beam at which electrons are accepted by the energy analyser. A deflection analyser that accepts electrons only at right angles to the photon beam gives a spectrum with different relative band areas from that given by a spherical retarding-field analyser, which accepts almost all the electrons. For the interpretation of spectra by using relative band intensities, the intensity must be integrated over all angles, so while results from spherical retarding-field analysers can be used directly, those from most differential analysers must first be corrected. It will be seen that the corrections are fortunately small in most instances, but on the other hand the measured angular distributions contain important information about the photoionization processes that can also be of help in the analysis of photoelectron spectra.

3.4.1 FORM OF THE ANGULAR DISTRIBUTION

The motion of the electrons ejected in photoionization can be described by wave-functions according to the angular momentum that the electrons must carry in order to satisfy the dipole selection rules. The angular distribution of the photoelectrons is determined by the s, p, d or f character of those outgoing spherical waves. The full form of the distributions will not be derived here, but can perhaps be made plausible by comparison with hydrogenic wave-functions. In ionization of an s electron from an atom, the electron wave must be a p wave, just as electric dipole-allowed transitions from S states lead only to P states in atomic spectroscopy. As the s electron wave-function has no defined orientation, the axis of the p wave-function is the direction defined by the dipole interaction, which is the direction of the electric vector of the electromagnetic wave. For plane-polarized light, the electric vector lies in the plane of polarization at right angles to the direction of propagation, so the electron signal can be measured as a function of the angle from the electric vector. The angular part of the electron wave-function is the same as that of an atomic p_z electron, where the electric vector of the light defies the z axis. The angular part of the p_z wave-function is the spherical harmonic Y_{10}, and the probability of observing an electron is proportional to its square. Hence, for the angular distribution $I(\theta)$:

$$I(\theta) \propto Y_{10}^2 = \left(\frac{3}{2\sqrt{\pi}}\cos\theta\right)^2 = \frac{3}{4\pi}\cos^2\theta \qquad (3.10)$$

In ionization of an atomic s orbital, the electrons therefore have a $\cos^2\theta$ distribution about the direction of the electric vector. When ionization is from a p orbital, on the other hand, s and d outgoing waves are allowed and usually both types of wave are involved. The s waves, like s atomic orbitals, are spherically symmetrical and correspond to isotropic angular distributions, while d wave distributions are peaked along the electric vector, but not so sharply as the p wave distributions.

Whatever the mixture of s, p, d or higher partial waves involved, the photoelectron angular distributions can be expressed by a single equation. If the angle of observation is measured from the direction of the electric vector of a plane-polarized light beam, the equation is

$$I(\theta) = \frac{\sigma}{4\pi}\left[1+\frac{\beta}{2}\left(3\cos^2\theta-1\right)\right] \qquad (3.11)$$

where σ is the total cross-section integrated over all angles and β, called the anisotropy parameter, is the single parameter needed to characterize the photoelectron angular distribution. For a pure p wave, β has the value $+2$, and equation 3.11 reduces to equation 3.10 apart from the inclusion of σ. In most photoelectron spectrometers the light is unpolarized, and the intensity must be measured as a function of the angle θ' away from the direction of the light beam. The distribution of intensity is then ·

$$I(\theta') = \frac{\sigma}{4\pi}\left[1+\frac{\beta}{2}\left(\frac{3}{2}\sin^2\theta'-1\right)\right] \qquad (3.12)$$

where β is the same parameter as in equation 3.11. The possible range of β values is from -1 to $+2$, and the value of β completely determines the angular distributions in the ionization of both atoms and molecules.

Proofs that equations 3.11 and 3.12 are valid for all atoms[16], for diatomic molecules[17] and for all molecules[18] have been given only recently. Relationships between β and theoretical quantities that describe the ionization process have been derived for atoms[16] and diatomic molecules[17]; they are complicated and involve several unknowns. The formula for β in atomic ionization illustrates the factors involved:

$$\beta =$$

$$\frac{l(l-1)\sigma_{l-1}^2+(l+1)(l+2)\sigma_{l+1}^2-6l(l+1)\sigma_{l+1}\,\sigma_{l-1}\cos(\delta_{l+1}-\delta_{l-1})}{(2l+1)(l\sigma_{l-1}^2+(l+1)\sigma_{l+1}^2)}$$

$$(3.13)$$

where l is the angular momentum of the electron in the atom before ionization, where Russell–Saunders coupling is assumed, σ_{l-1} and σ_{l+1} are the partial cross-sections for production of the $l-1$ and $l+1$ waves, and $(\delta_{l+1}-\delta_{l-1})$ is the phase difference between the two waves. The value of β depends not only on the strengths of the two partial waves but also on their phases, which control the interference between them. Negative β values can arise only when the interference term in equation 3.13 is large. Both the partial cross-sections and the phases depend on the electron energy, so β values are energy dependent.

Because of the mathematical difficulties, direct calculation of β values has been attempted only for very few atoms and molecules[19-21]. Some qualitative generalizations can be made, however, both from the theory and from the results of experiment[22], and these generalizations are helpful in the analysis of photoelectron spectra.

(1) The β values depend almost entirely on the nature of the orbital from which the electrons come, and not on details of the ionic states produced. Within the vibrational structure of a single band, β is constant, and if two or more bands result from ionization out of a single orbital (e.g., because of spin–orbit coupling or Jahn–Teller effects), the β values for such bands are the same. This relationship probably holds also for the several bands per orbital produced by ionization of open-shell molecules, unless the electron energies are very different.

(2) Theory predicts[19] and experiment has confirmed[23] that in the energy range covered by ultraviolet photoelectron spectroscopy, β values are energy dependent. They approach asymptotic values when the ejected electrons have energies of several hundred electron volts. The variation is especially rapid near the threshold, where the electron energy is very low, and therefore the angular distributions of electrons with energies below about 5 eV may not be characteristic of the orbitals ionized.

(3) Just as ionization out of a pure s orbital of an atom gives $\beta = 2$, so ionization from molecular orbitals built up mainly from atomic s orbitals is likely to give relatively high β values. In $1s\sigma_g^{-1}$ ionization of hydrogen, for instance, $\beta = 1.8$ and in $2s\sigma_u^{-1}$ ionization of nitrogen, $\beta = 1.25$. It has been found that in the ionization of atoms out of orbitals with l greater than one, high l values are associated with low β values, and the higher is l the lower is β. This relationship can be generalized to molecular ionizations if a theoretical means is found of estimating l in molecular orbitals, where it is not a good quantum number.

(4) Autoionization can seriously affect the angular distributions. If a molecular excited state has a lifetime before autoionization comparable with a rotation period, memory of the original direction of the electronic transition moment will be lost

and the ejected electrons will have an isotropic distribution ($\beta = 0$). Even when this is not so, the angular distributions of autoionization electrons will not generally be the same as those characteristic of direct ionization. Bands in photo-electron spectra to which autoionization contributes may have anomalous β values, and as the vibrational structure is also different in autoionization, β may vary within a band. The clearest example of this effect is the autoionization of oxygen at 736 Å (*Figure 3.3*); the first few peaks in the $^2\Pi_g$ band, to which the contribution from direct ionization is large, have $\beta \approx -0.5$, whereas the later peaks up to $v' = 12$ caused by autoionization alone have $\beta \approx 0.0$.

3.4.2   EXPERIMENTAL METHODS AND EXAMPLES

The basic experimental method of investigating angular distributions of photoelectrons is to bring about photoionization in a small defined volume and to measure the electron signals at different angles by moving either the analyser or the light beam. There are many variants of this method, which has been used to establish the validity of equations 3.11 and 3.12 experimentally, and to measure the most accurate β values. A schematic diagram of one form of suitable apparatus is shown in *Figure 3.6*, and *Figure 3.7* illustrates the form of the experimental angular distributions.

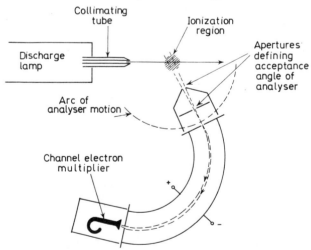

Figure 3.6. Schematic diagram of an apparatus for measuring the angular distribution of photoelectrons

When the ionization volume is defined by the position of a narrow light beam passing through a diffuse gas, the effective volume from which photoelectrons can enter the analyser is a function of the angle of measurement. The raw experimental data must usually be multiplied by $\sin \theta'$ in order to correct for this volume effect

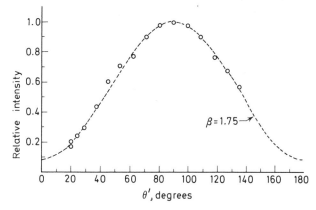

Figure 3.7. Experimental angular distribution of photoelectrons (corrected by multiplication by $\sin \theta'$) from ionization of hydrogen by HeI light to the state $X^2\Sigma_g^+, v' = 2$ of H_2^+. (From Carlson, T. A. and Jonas, A. E., *J. chem. Phys.*, **55**, 4913 (1971), by courtesy of the American Institute of Physics)

and obtain the true angular distribution. The anisotropy parameter can be evaluated from the experimental angular distribution, $I(\theta')$, by making a least-squares fit to the theoretical distribution in the form

$$I(\theta') = 1 + B \sin^2 \theta' \qquad (3.14)$$

where

$$B = \frac{3\beta}{4 - 2\beta} \qquad (3.15)$$

Because of the simple form of the angular distributions, the value of β can be found by measuring the intensity at just two angles. The angle 54 degrees 44 minutes [54.736 degrees or $\cos^{-1} (1/\sqrt{3})$], the so-called 'magic' angle, is particularly useful, because at this angle the intensity is completely independent of β whether the light is polarized or not. It is very convenient to choose this angle as one of those at which photoelectron spectra are determined, as relative intensities in the spectrum are here undistorted by variations in angular distributions. Measurements at just two angles suffice in any event to determine β, and ingenious devices have been proposed in order to accomplish this determination without physical motions in simple photoelectron spectrometers[24]. The anisotropy parameters for some important photoelectron bands are given in *Table 3.1* in order to illustrate the range of values encountered. One example of the variation of β within a single band is included, namely the $2p\sigma_g^{-1}$ ionization of nitrogen by He I light to give N_2^+ ($X\,^2\Sigma_g^+$) ions. The marked deviation of the β value for $v' = 1$ from the values for $v' = 0$ and $v' = 2$ is difficult to interpret: although it might be due to autoionization at 584 Å, there is little supporting evidence for this autoionization. The measured vibrational line intensities are well represented by Franck–Condon factors calculated for direct ionization, so the contribution of autoionization to the vibrational structure must be small. An alternative explanation would be a breakdown of the Born–Oppenheimer approximation, a coupling between vibrational and

Table 3.1   ANISOTROPY PARAMETERS

Substance	Ionic state	Photon energy, eV	β
Ar	$^2P_{\frac{3}{2}}, ^2P_{\frac{1}{2}}$	21.22	0.85 ±0.05
Kr	$^2P_{\frac{3}{2}}, ^2P_{\frac{1}{2}}$	21.22	1.20 ±0.05
Xe	$^2P_{\frac{3}{2}}$	21.22	1.45 ±0.05
	$^2P_{\frac{1}{2}}$	21.22	1.35 ±0.05
	$^2P_{\frac{3}{2}}$	16.85	1.28 ±0.05
	$^2P_{\frac{1}{2}}$	16.85	1.15 ±0.05
H_2	$X\,^2\Sigma_g^+$	21.22	1.75 ±0.05
N_2	$X\,^2\Sigma_g^+$ ($v'=0$)	21.22	0.69 ±0.03*
	$X\,^2\Sigma_g^+$ ($v'=1$)	21.22	1.37 ±0.03*
	$X\,^2\Sigma_g^+$ ($v'=2$)	21.22	0.6 ±0.1 *
	$A\,^2\Pi_u$	21.22	0.46 ±0.05*
O_2	$X\,^2\Pi_g$	21.22	−0.13 ±0.03*
	$X\,^2\Pi_g$ ($v'=1$)	16.85	−0.53 ±0.03*
CO_2	$X\,^2\Pi_g$	21.22	−0.2 ±0.1
	$A\,^2\Sigma_u$	21.22	0.7 ±0.1
	$B\,^2\Sigma_u^+$	21.22	−0.5 ±0.1
	$C\,^2\Sigma_g^+$ ($v'=0$)	21.22	1.2 ±0.2
H_2O	$^2B_1$	21.22	1.0 ±0.1
	$^2A_1$	21.22	0.3 ±0.1
	$^2B_2$	21.22	−0.1 ±0.2
CH_3I	$^2E_{\frac{3}{2}}, ^2E_{\frac{1}{2}}$	21.22	1.5 ±0.1
	$^2A_1$	21.22	0.6 ±0.2
	^2E	21.22	0.9 ±0.2

Data in this table are taken from the work of Carlson *et al.*[22] and from Carlson, T. A. and Jonas, A. E., *J. chem Phys.*, **55**, 4913 (1971), and those marked with an asterisk from the results of R. Morgenstern, A. Niehaus and M. W. Ruf presented at the VII IPCEAC Conference, Amsterdam, 1971. The results from this last group generally deviate by +0.1 unit in β from those of Carlson's group.

electronic motions. If this were true, one would expect variations in β values in ionization processes where there is definitely an interaction of nuclear and electronic motion, as for instance in ionization to states that undergo a Jahn–Teller effect. Although bands that correspond to such states have been examined (e.g., the first band in the spectrum of methane), no variation of β within them has been found. In addition to this case of N_2^+, anomalous β variations within resolved vibrational structure have been found[22] in the analogous band for $X\ ^2\Sigma^+$ of CO^+ and in the fourth bands in the spectra of the triatomic molecules CO_2 and COS.

The β values given in *Table 3.1* for ionizations by He I light are all (except the $1s\sigma^{-1}$ ionization of H_2) within the range from -0.5 to $+1.5$, and this range is typical of the anisotropies that have been measured. In a photoelectron spectrometer that accepts photo-electrons in a narrow angle around 90 degrees to the photon beam, this range corresponds to a maximum variation in relative band intensities by a factor of 1.6 due to variations in angular distribution alone. This factor is small enough not to affect seriously the applica-tion of the limiting rules for band intensities given in Chapter 1, Section 1.3. Within the photoelectron spectrum of a single compound, the range of variation of β values is usually even smaller than this, especially if bands at very low electron energy are excluded from consideration.

3.4.3 ELECTRON SPIN AND MOLECULAR ROTATION

In considering the selection rules for photoionization in Section 3.2, we neglected, or rather took an average over, two sources of angular momentum that may affect angular distributions, namely the spin of the photoelectrons and the rotation of the molecules. The spin of the outgoing electron can have no effect on the measured angular distributions unless the ionizing light is circularly polarized and the electron detector is sensitive to electron spin polarization[18]. When both of these conditions are fulfilled, a new equation must be used instead of equation 3.11 or 3.12, but the new equation reduces to these two, however, in simple cases. It can happen that photoelectrons ejected in the propagation direction of a circularly polarized light beam are fully spin-polarized, and this effect has been observed in ionizations of alkali metal atoms[25]. This effect may find applications as a source of spin-polarized electrons, but its use as an aid to the analysis of photoelectron spectra seems remote.

The theory of the partial cross-sections and electron angular distributions for individual rotational transitions in the photoionization of diatomic molecules has been developed by Bucking-ham, Orr and Sichel[17]. It was shown that equations 3.11 and 3.12 are valid for the individual rotational lines and that the angular distributions for transitions with different ΔJ values will, in general, differ. The theory has been applied in detail, however, only to the derivation of the partial cross-sections in the one instance of ionization of hydrogen, which is the only molecule in which the rotational quanta are sufficiently large to be resolved in photo-electron spectroscopy. Åsbrink[26] measured the photoelectron spectrum of hydrogen excited by Ne I light and attained a resolution of 4 meV, with which he was able partially to resolve the rotational structure. Hydrogen is a special case, not only because of the large rotational intervals, but also because both H_2 and H_2^+ exist in *ortho* and *para* states in which the spins of the two nuclei are parallel and antiparallel, respectively. Because the total wave-function must remain antisymmetrical for exchange of the nuclei (Fermions), *ortho*-hydrogen exists only in states with odd rotational quantum numbers, J, and *para*-hydrogen only in states with zero or even values of J. The same arrangement holds for H_2^+ as for H_2, and as the nuclear spins are not affected in an electronic transition there is a strict selection rule $\Delta J = 0, \pm 2, \pm 4 \ldots$. This rule, or a related rule, also holds in the ionization of any molecule with identical nuclei in equivalent positions, e.g., O_2, CO_2, C_2H_2 and NH_3. The rule was used by Sichel[27], who applied the general theory of Buckingham, Orr and Sichel[17] to the photoionization of hydrogen. The work of Åsbrink, and also the measurements of Brehm and Frey[28] on the photoionization of hydrogen by He I light, showed that the transitions with $\Delta J = 2$ are much less intense than those with $\Delta J = 0$. According to the theory, this result indicates that the wave-function for the H_2 molecule deviates little from spherical symmetry. In a later and most elegant experiment, Niehaus and Ruf[29] measured the angular distributions of the photoelectrons from hydrogen ionized by Ne I radiation. They obtained separate β values for the different rotational transitions, namely for $\Delta J = 0$, $\beta = 1.95 \pm 0.3$ and for $\Delta J = 2$, $\beta = 0.85 \pm 0.14$. When $\Delta J = 0$, the value of β is so near to $+2$ that the outgoing electron must be almost a pure p wave. The very different angular distribution for $\Delta J = 2$ clearly shows the importance of the rotational angular momentum, but no clear theoretical explanation for the observed β value has yet been given. According to Sichel's theory[27], the outgoing electron wave will be an admixture of p and f waves with interference between them, which must lead to a value of β less than 2, but this value has not, however, been calculated explicitly. In a more recent theoretical formulation, Dill[30] started from the assumption that the outgoing electron wave is always a p wave and obtained $\beta = 0.2$ for $\Delta J = 2$, but also pointed out that the experimental value indicates that f wave participation is important.

It is interesting that the experimental value of $\beta = 0.85$ for $\Delta J = 2$ agrees very well with a pure f wave, for which $\beta = 0.80$. This result would correspond to a simple addition of the angular momentum changes in rotational and electronic motion in determining the angular momentum of the outgoing electron, and to a zero cross-section for the p wave contribution. It is not clear whether this correspondence is more than a coincidence, however, as the value of $\beta = 0.8$ might be the fortuitous result of interference between p and f waves of appropriate phases and amplitude.

3.5 OTHER TECHNIQUES

The study of positive molecular ions by techniques other than photoelectron spectroscopy has a long history, and in its early stages the development of photoelectron spectroscopy was greatly assisted by the knowledge of ionic states that was already available. Reliable first ionization potentials of atoms and molecules were obtained by the study of Rydberg series in absorption spectra, and also by the conventional technique of photoionization. Some excited ionic states had been discovered as limits of Rydberg series, and others were known from the emission spectra of the ionized species. All of these techniques have taken on a new lease of life in recent years, partly as a result of the stimulus of photoelectron spectroscopy, and new techniques have been added to them.

3.5.1 EMISSION SPECTRA

Excited states of positive ions can be created not only by photo-ionization but also by impact with electrons, ions or even neutral species, and all of these processes take place in a gas discharge. The excited ions can emit light in reverting to a lower ionic state, and so produce emission spectra. Because these fluorescence transitions occur between two bound states of the ion, they are governed by the selection rules for dipole radiation. Details of the emission spectrum, especially the rotational fine structure, can indicate the electronic species of the two states involved and also the molecular geometry of the ions in those states. The origin of an emission band gives the energy difference between two ionic states very precisely, and any broadening of the lines in a band may indicate something about the pre-dissociations undergone by the ions. This information about ions from emission spectra is sometimes much more detailed than that which can be obtained from a photoelectron spectrum, but unfortunately it is very rarely available. Whereas the emission spectra of atomic ions are commonplace, emission spectra of only very few molecular ions have been observed. One reason for this is that many molecular ions in excited states dissociate before they can emit light; another reason may be that although ionic emission bands are present, they have not been recognized as such. A knowledge of the positions of ionic states from photoelectron spectroscopy permits the prediction of the possible positions of ionic emission bands, which should make their detection easier.

3.5.2 ENERGY-LOSS SPECTRA

The relevance of the study of Rydberg states to the analysis of photoelectron spectra has been pointed out already in Chapter 1. Until recently, these states could be detected only by the extremely difficult technique of vacuum ultraviolet absorption spectroscopy, and because of the difficulties involved few molecules had been studied. This situation has changed with the development of the new technique of electron energy-loss spectroscopy[31, 32].

An electron beam of defined energy is passed through a cell containing a gas, and the electrons that are scattered inelastically in the direction of the primary beam are examined in an electron energy analyser. The primary electrons can lose energy by exciting the gas molecules rotationally, vibrationally or electronically in the process

$$A + e \rightarrow A^* + e \tag{3.16}$$

For the kinetic energies of the electrons before and after scattering, KE_1 and KE_2, respectively, we have

$$KE_2 = KE_1 - E^* \tag{3.17}$$

The energy lost by the electrons, $KE_1 - KE_2$, is exactly equal to the energy gained by the molecules, and must correspond to the energy difference between the ground state and an excited state. The energy-loss spectrum therefore contains the same sort of information as an optical absorption spectrum but can cover the whole energy range from zero to hundreds of electronvolts with no special difficulty.

Another advantage of the method is its sensitivity, which can exceed that of conventional vacuum ultraviolet spectroscopy by several orders of magnitude. The resolution is not so high, however, as that which can be attained in optical spectroscopy. The selection rules for excitation by electron impact are not necessarily the same as those that govern optical transitions, and the differences are most apparent at low primary electron energies. When electrons scattered along the primary beam direction are studied, the relative band intensities are proportional to optical transition probabilities for primary electron energies of 300 eV or more. At lower electron energies and for electrons scattered through large angles, the selection rules are relaxed, and singlet–triplet transitions, for instance, may give intense energy-loss bands. Energy-loss spectroscopy therefore enables one to detect both optically allowed and

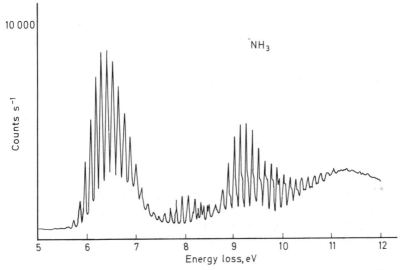

Figure 3.8. Energy-loss spectrum of ammonia. (From Rendina and Grojean[32], by courtesy of the Society for Applied Spectroscopy)

optically forbidden transitions, and to distinguish between them by the energy dependence of their relative intensities or by the angular distribution of the scattered electrons. All of these characteristics can be exploited in order to study Rydberg states in conjunction with photoelectron spectra. The vibrational structure of electronic excitation bands in energy-loss spectroscopy is governed by the Franck–Condon principle in the same way as photoionization or absorption, so the Rydberg bands resemble the photoelectron bands on which they converge. An example is given in *Figure 3.8*, which shows the energy-loss spectrum of ammonia taken on a commercial spectrometer.

Electron impact can, of course, cause ionization as well as excitation. In a direct electron impact ionization process:

$$A + e \rightarrow A^+ + 2e \qquad (3.18)$$

two electrons leave the collision complex and the excess energy can be distributed between them. The electron energy distribution is therefore continuous and does not interfere with the observation of discrete energy losses in the excitation process of equation 3.16. When the energy of the primary electrons is well above the ionization limit, the electron energy distribution for ionization has two broad peaks, one near the primary beam energy and the other nearer zero energy. One of the electrons can be thought of as a primary electron that has lost energy, and the other as a secondary electron. The energy loss of the primary electron is equivalent to the photon energy in photoelectron spectroscopy, and the secondary electron to the photoelectron. The secondary electron spectrum for a single energy loss is equivalent to a photoelectron spectrum, but as the primary energy losses are continuously distributed no structure that resembles a photoelectron spectrum is normally visible. This difficulty can be overcome in a coincidence experiment in which primary electrons that have lost energy are detected in one electron energy analyser and secondary electrons in another. By using the coincidence technique, it is possible to detect the particular secondary electrons that correspond to a chosen primary electron energy loss, using the fact that the two electrons must leave the collision simultaneously. The coincidence experiment is equivalent to photoelectron spectroscopy with a completely tunable light source, as any primary energy loss can be chosen. This experiment is technically very difficult, but some success has already been achieved[33].

3.5.3   NEW PHOTOIONIZATION TECHNIQUES

When photons impinge upon molecules, they can give rise to a photocurrent not only by direct ionization and autoionization but also by ion-pair production:

$$AB + h\nu \rightarrow A^+ + B^- \qquad (3.19)$$

In order to distinguish this process from photoionization, it is necessary to analyse the ions according to their masses and charges in a mass spectrometer, and the combined technique is called photoionization mass spectrometry. The mass spectra also show the dissociations that the excited molecular ions produced in photoionization undergo, and this topic is discussed in Chapter 7.

The addition of a mass spectrometer does not make it possible, however, to distinguish between autoionization and direct photo-ionization, and for this purpose a new technique has been developed. In photoionization resonance spectroscopy[34, 35], sometimes called steradiancy analysis, the photon energy is varied and the signal recorded represents the number of photoelectrons formed with zero kinetic energy. Photoelectrons can be ejected with zero energy only when the energy of the photons matches exactly that of an ionic state, so this is a true threshold process. Autoionization which cannot yield electrons with zero kinetic energy is excluded. The spectrum produced has the form of a differential photoelectron spectrum, and contains a peak at each energy where an ionic state is produced. The relative intensities of electronic bands and vibrational lines are slightly different from those found in photoelectron spectroscopy, because the fall-off of ionization cross-section above the threshold is eliminated and the angular distributions of the photoelectrons also have no effect. A most important advantage of the technique is that the energy scale is established absolutely by the photon energy and is independent of the contact potential variations that are so troublesome in photoelectron spectroscopy. The Doppler effect of motion of the target molecules on resolution is also eliminated, because the electrons have zero energy (equation 2.14). Although the intensity of light sources that emit a continuous range of wavelengths is much lower than that of resonance lamps, the sensitivity for the detection of electrons with zero energy can be higher by several orders of magnitude than that for electrons with higher energy. Photoionization resonance spectroscopy is therefore a technique very closely related to photoelectron spectroscopy.

3.5.4 PENNING IONIZATION

Penning ionization[36, 37] is brought about not by photons but by metastable atoms, usually rare gas atoms:

$$A + B^* \rightarrow A^+ + B + e \qquad (3.20)$$

The process resembles photoionization in that only one electron is involved, and because of its low mass this electron carries away most of the excess energy of the process. The spectrum of Penning electrons resembles a photoelectron spectrum, but also differs in the following important respects. The available energy is the excitation energy of B* plus a contribution from the relative kinetic energies of A and B*, and this energy provides for ionization and excitation of A^+, as in photoelectron spectroscopy. Part of the excess energy after the collision goes into relative translation of A^+ and B, although most appears as electron energy. This complication of the energy balance has the effect that peaks in Penning electron spectra are broader than those in photoelectron spectra, even when the original particles have only thermal kinetic energies. The peaks are also shifted by the heavy-particle kinetic energy effects, so that the precise energies of the ionic states of A^+ produced cannot be determined. The exact energies of the ionic states are usually taken from photoelectron spectroscopy, whereupon the measured shifts give information about the details of the Penning ionization process itself[38].

The transition probabilities for producing different electronic states of A^+ are different in Penning ionization and photoionization, and in Penning ionization they also depend on the nature of the excited atom B*. There is nothing surprising about this effect as the interaction processes are different, but the differences are interesting and could be useful. One likely possibility is that two-electron processes may be more intense in Penning ionization than in photoionization; another is that the differences between Penning ionization and photoionization, particularly when He* ($2\,^3S$) metastables are used as B*, may be indicative of the nature of the orbitals from which electrons are ionized. There is some evidence that large antibonding orbitals have higher cross-sections relative to other orbitals in Penning ionization than in photoionization, so a comparison could, in principle, assist in the interpretation of photoelectron spectra. This comparison seems to be impractical at present because of the technical difficulties involved in the Penning ionization experiments, but it is a possibility for the future.

3.5.5   ELECTRON IMPACT IONIZATION

The traditional method of measuring ionization potentials by electron impact is to record the molecular ion current in a mass spectrometer while increasing the energy of the thermionic electron beam used for ionization. When the electron energy is equal to the ionization potential, the cross-section for electron impact ionization is zero, and it should increase linearly with increasing energy above the threshold. Ionization potential differences between the sample and a standard are obtained by various methods of extrapolation or other treatment of the measured ion currents. In fact, the energy spread of thermionic electrons from a heated filament is so broad, and the interference of autoionization processes with the form of the cross-section behaviour so serious, that not even first ionization potentials can be measured accurately or reliably by this method. Mono-energetic electron beams can now be used instead of thermionic electrons, which makes it possible to see some structure in the cross-section functions. It is no easier to find inner ionization potentials by electron impact methods than by photoionization, however; because the cross-sections are zero at the threshold, it is even more difficult. It may nevertheless be interesting to compare electron impact ionization cross-sections with photoionization cross-sections and photoelectron spectra, as the selection rules for electron impact ionization near the threshold are different from those that govern photoionization. Ionic states that are not observed in photoelectron spectroscopy may be accessible by electron impact ionization, particularly those states which are reached by transitions with ΔS greater than unity.

REFERENCES

1. CONDON, E. U. and SHORTLEY, G. H., *The Theory of Atomic Spectra*, Cambridge University Press (1935)
2. ORCHARD, A. F., *Discuss. Faraday Soc.*, **54**, 252 (1973)
3. SAMSON, J. A. R., *Chem. Phys. Lett.*, **12**, 625 (1972)
4. KRAUSE, M. O., *Phys. Rev.*, **177**, 151 (1969)
5. CHANG, T. N., ISHIHARA, T. and POE, R. T., *Phys. Rev. Lett.*, **27**, 838 (1971)
6. SAMSON, J. A. R. and CAIRNS, R. B., *Phys. Rev.*, **173**, 80 (1968)
7. BLAKE, A. J., *Proc. R. Soc., Lond.*, **A325**, 555 (1971)
8. BLAKE, A. J. and CARVER, J. H., *J. Quant. Spec. Rad. Transfer*, **12**, 59 (1972)
9. CARVER, J. H. and GARDNER, J. L., *J. Quant. Spec. Rad. Transfer*, **12**, 207 (1972)
10. BLAKE, A. J. and CARVER, J. H., *J. chem. Phys.*, **47**, 1038 (1967)
11. BERRY, R. S., *J. chem. Phys.*, **45**, 1228 (1966)
12. BERKOWITZ, J. and CHUPKA, W. A., *J. chem. Phys.*, **51**, 2341 (1969)
13. BLAKE, A. J., BAHR, J. L., CARVER, J. H. and KUMAR, V., *Phil. Trans. R. Soc., Lond.*, **A268**, 159 (1970)
14. KEIMENOV, V. I., CHIZHOV, YU. V. and VILESOV, F. I., *Optics Spectrosc.*, **32**, 371 (1972)
15. SMITH, A. L., *Phil. Trans. R. Soc., Lond.*, **A268**, 169 (1970)
16. COOPER, J. and ZARE, R. N., *J. chem. Phys.*, **48**, 942 (1968)
17. BUCKINGHAM, A. D., ORR, B. J. and SICHEL, J. M., *Phil. Trans. R. Soc., Lond.*, **A268**, 147 (1970)
18. BREHM, B., *Z. Physik*, **242**, 195 (1971)
19. KENNEDY, D. J. and MANSON, S. T., *Phys. Rev.*, **A5**, 227 (1972)
20. TULLY, J. C., BERRY, R. S. and DALTON, B. J., *Phys. Rev.*, **176**, 95 (1968)
21. SCHNEIDER, B. and BERRY, R. S., *Phys. Rev.*, **182**, 141 (1969)
22. CARLSON, T. A., MCGUIRE, G. E., JONAS, A. E., CHENG, K. L., ANDERSON, C. P., LU, C. C. and PULLEN, B. P., in Shirley, D. A. (Editor) *Electron Spectroscopy*, North Holland, Amsterdam, 207 (1972)
23. NIEHAUS, A. and RUF, M. W., *Z. Physik*, **252**, 84 (1972)
24. AMES, D. L., MAIER, J. P., WATT, F. and TURNER, D. W., *Discuss. Faraday Soc.*, **54**, 277 (1973)
25. HEINZMANN, U., KESSLER, J. and LORENZ, J., *Phys. Rev. Lett.*, **25**, 1325 (1970)

26. ÅSBRINK, L., *Chem. Phys. Lett.*, **7**, 549 (1970)
27. SICHEL, J. M., *Molec. Phys.*, **18**, 95 (1970)
28. BREHM, B. and FREY, R., *Z. Naturforsch.*, **26a**, 523 (1971)
29. NIEHAUS, A. and RUF, M. W., *Chem. Phys. Lett.*, **11**, 55 (1971)
30. DILL, D., *Phys. Rev.*, **A6**, 160 (1972)
31. LASSETRE, E. N., SKERBELE, A., DILLON, M. A. and ROSS, K. J., *J. chem. Phys.*, **48**, 5066 (1968)
32. RENDINA, J. F. and GROJEAN, R. E., *Appl. Spectrosc.*, **25**, 24 (1971)
33. MCDOWELL, C. A. *Discuss. Faraday Soc.*, **54**, 297 (1973)
34. VILLAREJO, D., HERM, R. R. and INGHRAM, M. G., *J. chem. Phys.*, **46**, 4994 (1967)
35. SPOHR, R., GUYON, P. M., CHUPKA, W. A. and BERKOWITZ, J., *Rev. scient. Instrum.*, **42**, 1872 (1971)
36. ČERMAK, V., *J. chem. Phys.*, **44**, 3774 (1966)
37. FUCHS, V. and NIEHAUS, A., *Phys. Rev. Lett.*, **21**, 1136 (1968)
38. HOTOP, H. and NIEHAUS, A., *Z. Physik*, **228**, 68 (1969)

4 Electronic Energies of Ionic States

4.1 INTRODUCTION

Photoelectron spectra provide a wealth of information on the energies of the electronic states of positive ions, which are closely related to the energies of molecular orbitals in neutral molecules. This is a major reason for the interest of theoreticians in photoelectron spectroscopy and has inspired many calculations of molecular electronic structures. In this chapter, the meaning of the experimental results and theoretical models are considered first, and then comparisons between experiment and calculation are discussed. These comparisons and the calculations themselves are most often made as an aid to the analysis of experimental photoelectron spectra, and here they are discussed mainly from this standpoint. Experimental energy information alone can sometimes be used in a direct way to make deductions about one aspect of molecular electronic structure, the atomic charge distribution, and this topic is examined in the final section. Throughout the chapter, the energies of orbitals and the energies involved in ionization are referred to, and when numerical values are given they are in electron-volts, units of energy. In photoelectron spectroscopy, the measured quantity is actually an electrical potential in volts and the electron energies are strictly derived quantities, although identical in magnitude. The phrase 'ionization energy' is generally used in this chapter and elsewhere when energies are being compared, and the more familiar 'ionization potential' is used to refer to experimental measurements. The two expressions can, in practice, be taken as synonymous without any change in sense.

4.2 ENERGIES FROM PHOTOELECTRON SPECTRA

In contrast to all other methods of investigating molecular ionization, photoelectron spectroscopy often gives both *adiabatic* and *vertical* ionization potentials. An adiabatic ionization energy is unambiguously defined as the difference in energy between the neutral molecule in its electronic, vibrational and rotational ground state and the ion in the lowest vibrational and rotational level of a particular electronic state. Adiabatic transitions are often seen in photoelectron spectra as the zero vibrational levels of each electronic state, that is, as the first vibrational lines in the different bands. The adiabatic transitions are weak, however, whenever there is a large change in equilibrium molecular geometry on ionization. Even in the spectra of triatomic molecules there are bands in which the true adiabatic ionization potentials may not yet have been found, for example, in the $^2A_1$ states of H_2O^+ and H_2S^+. In interpreting the spectra of polyatomic molecules, it is unwise to call the onset of any broad band an adiabatic ionization potential unless the shape of the band definitely indicates that the (0–0) transition has been seen. For many molecules with non-rigid skeletons, such as the aliphatic hydrocarbons, the adiabatic values are simply not known for this reason.

The vertical ionization energy is defined as the energy difference between the molecule in its ground state and the ion in a particular electronic state, but with the nuclei in the same positions as they had in the neutral molecule. The vertical transition therefore corresponds to a vertical line through the molecular ground state on a potential energy diagram, and it is usually defined experimentally as the point of maximum intensity in a photoelectron band. These definitions both lack rigour. It is not clear, for instance, whether the nuclear positions in the molecular ground state should be defined by the maximum of the vibrational wave-function, the maximum of its square or the minimum of the electronic surface. In a resolved band, the vertical transition may not exactly match a vibrational level, and then the experimental value is ambiguous. Brehm[1] has proposed that the vertical ionization potential be defined experimentally as the centre of gravity of a band. This definition has the advantage of a clear theoretical interpretation in terms of potential energy surfaces, but at the expense of simplicity. The advantage that the maximum of a band can easily be located experimentally even when bands overlap makes the vertical ionization energy defined thereby the most generally useful experimental measure of ionization energy.

Whether defined as adiabatic or vertical, the experimental ionization energies are the differences in total energy between the neutral molecule and the ion in a particular state:

$$I = E(A^{+*}) - E(A) \qquad (4.1)$$

This exact expression suggests that the correct way to calculate ionization energies is to determine the total energy of the ion and of the molecule separately, and then subtract the two values. When this is done, the energy of the ion is sometimes calculated at its own equilibrium geometry instead of at that of the molecule; the result must then be compared with an adiabatic ionization energy. Much more commonly, Koopmans' approximation is used and an orbital energy is available for comparison with experiment. Koopmans' theorem involves no allowance for the reorganization of the nuclei or of the electrons, and therefore it seems logical to compare orbital energies with vertical ionization energies derived from experiment.

4.3   MOLECULAR ORBITALS, ORBITAL ENERGIES AND KOOPMANS' THEOREM

Koopmans' theorem defines molecular orbital energies as the difference in energy between an electron at an infinite distance from the molecular ion and the same electron in the molecule. Unfortunately, this definition is really one of the ionization energy and is valid for orbital energy only within one approximate model of molecular electronic structure, the self-consistent-field (SCF) model, and then only with special assumptions. In the SCF model and in all models that give recognizable molecular orbitals, they are *one-electron* orbitals; the electrons are taken singly and treated as moving in a field produced by the fixed nuclei and the averaged interactions with the other electrons, including exchange interactions[2]. If the interactions of the remaining electrons remain exactly the same after one of their number has been removed, then Koopmans' theorem holds for SCF orbital energies. Furthermore, although the concept of orbitals seems to be clear from long familiarity, the only unambiguous definition of orbitals and their energies is a mathematical definition based on the SCF model. They must be defined, tedious as it is, as the eigen-vectors and eigen-values of an SCF Hamiltonian. Although these definitions are precise, there are still two important difficulties in the idea of using molecular orbitals as a model for the electronic structure and, hence the photoelectron spectra of real molecules. Firstly, the motion and energy of an electron are not independent of the detailed positions and motions of the other electrons, particularly those in the same shell. The electrons, so far as their instantaneous positions have a meaning, tend to keep apart and so reduce their mutual repulsion energy; their motions are correlated. The only way to include this electron correlation in calculations for many-electron systems is to use the method of configuration interaction. One starts with an electron configuration with electrons in particular orbitals, corresponding to the ground state of the molecule, and then mixes in the wave-functions for excited states that have the same total symmetry but different electron configurations. The mixing coefficients are found by the variation method[2], and it can be proved that if an infinite number of excited configurations are properly mixed in, the electron correlation will be represented correctly. The result is a good total wave-function for all electrons and a good total energy, but individual molecular orbitals no longer have any meaning. In other words, the idea of one-electron molecular orbitals is an approximation from the outset.

The second difficulty arises from the nature of the one-electron orbitals themselves. They are solutions of a Schrödinger equation, and it is a general property of such solutions that linear combinations of them also satisfy the equation. Delocalized molecular orbitals can be converted by linear combination into, for instance, the equivalent orbitals of Pople[3], which represent localized bonding contributions. The new orbitals are just as good solutions of the Schrödinger equation as the previous orbitals, but they have different energies. The question arises of which set of orbitals and orbital energies one should choose in order to make a comparison with the photoelectron spectrum. The practical solution to this question, from experience in photoelectron spectroscopy, is that the fully delocalized orbitals must be chosen, but a choice on this basis will not satisfy a theoretician. The theoretical requirement is that the set of wave-functions must be chosen that gives the lowest energies under Koopmans' theorem. In her original paper, Koopmans[4] showed that the correct choice corresponds to the normal shell model of the atom or molecule, that is, to the delocalized orbitals, at least for the lowest ionic state of each symmetry. Extension to excited ionic states of each symmetry species is experimentally justified by the success of Koopmans' theorem for inner shells, and a theoretical justification has also been given[5].

We can now consider the approximations involved in practical calculations of ionization energies, which include the limitations of molecular orbital models in general, those of the exact model used in each instance, and of Koopmans' approximation itself[6]. The most important limitation of molecular orbitals is the neglect of correlation, which enters the calculation of an ionization energy as the difference in correlation energy between the neutral molecule and the ion; the neglect of correlation should, in general, make the calculated ionization energy too small. This will only be so in practice, however, if the model calculation is an SCF model carried to the Hartree–Fock limit (see Section 4.4) and made for both molecule and ion, because only with such a calculation is the neglect of correlation energy the one remaining serious approximation. In using Koopmans' theorem with a molecular orbital model, the further approximation that the electron interactions are exactly the same in the ion and molecule is introduced, which is manifestly not true. As the electrons in the ions can always attain a more stable state than the one defined by their motions in the molecule, the use of Koopmans' theorem should give an ionization energy that is too high. The difference between the ionization energy calculated by Koopmans' theorem and that obtained by calculating the total energies of the ion and molecule separately and then subtracting is called the reorganization energy, and both this and the correlation energy can be expected to vary from one ionic state to another. These are clearly two good reasons why even sophisticated calculations may not give correct ionization energies quantitatively; a more important question in photoelectron spectroscopy is whether the ordering of the ionic states can be calculated correctly. This depends on the magnitudes of deviations in the reorganization energy and the correlation energy from ionic state to state compared with the energetic separation of one state from another. It has to be investigated by comparison between theory and experiment, and this comparison is the subject of the next section. In general, it seems that the cancellation of errors in using Koopmans' theorem due to neglect of both the correlation energy and reorganization energy is only partial, as the reorganization energy is usually larger and results in calculated ionization potentials that are higher than the observed values. The variations from state to state are such that if experimental ionization potentials are separated by less than 1 eV, even the best molecular orbital calculations cannot be relied upon for their ordering.

All of the preceding remarks apply in the first instance to the ionization of closed-shell molecules; when open-shell molecules are ionized, Koopmans' theorem in its simple form does not apply[6]. A theoretical approach to the problem has been made by Dodds and McWeeney[7], who have developed a formalism within which an analogue of Koopmans' theorem is satisfactory. The experimental fact is that in the ionization of an open-shell molecule, several ionic states are often attained by the removal of electrons from the same molecular orbital. There is no longer a one-to-one correspondence between bands in the spectrum and orbitals in the molecule, so a practical problem exists of relating the several observed ionization potentials to a single orbital energy. Evans, Green and Jackson[8] have proposed that an orbital energy can be derived by taking a weighted mean of the ionization potentials for all of the bands that correspond to a particular electron configuration. The weights are the theoretical relative cross-sections for the different ionic states, equal to their total degeneracies or fractional parentage coefficients[9]. The mean ionization energies derived in this manner are related to orbital energies by the same approximations as those for closed-shell molecules.

4.4   MOLECULAR ORBITAL CALCULATIONS

It is usual to distinguish four types of molecular orbital calculations, and these types are listed below in order of decreasing difficulty and therefore increasing complexity of the molecules for which such calculations can be made:

(1) 'exact' Hartree–Fock calculations;
(2) *ab initio* SCF calculations;
(3) semi-empirical calculations;
(4) empirical calculations.

An exact' Hartree–Fock (HF) calculation means a calculation by the SCF method in which no terms are neglected and no further improvement of the total energy can be gained either by increasing the number of iterations or by expanding the basis set, the set of orbitals, atomic or otherwise, out of which molecular orbitals are built.

The advantage of the 'exact' HF method is that the meaning of the results is well defined. The calculations do not involve mathematical approximations or arbitrary parameters but only the approximations inherent in the method itself. These approximations are principally the neglect of electron correlation and of relativistic effects for inner shells. The direction in which these approximations will affect ionization energies can be clearly predicted, even though the exact magnitudes of the corrections are not known. If Koopmans' theorem is used to derive ionization energies, the reorganization energy approximation is also involved, but for HF orbitals this is of a definite direction and a magnitude that can be estimated. Ionization energies derived from HF calculations by Koopmans' theorem will definitely be too high, as reorganization energy is the dominant approximation, whereas those derived by re-calculation of the total electronic energies in the ionic states will be too low because the correlation energy is less in the ion than in the molecule. These predictions are fully borne out by the comparisons with photoelectron spectra. The calculations are unfortunately limited to relatively few small molecules simply because of their cost in terms of computer time.

Some of the advantages of the full HF calculation are shared by *ab initio* SCF calculations for larger molecules, which are generally made possible by mathematical simplification through contraction of the basis sets[10]. Although these are not true HF calculations, they do not involve any arbitrary parameters, so the effect of the approximations on the calculated energies remains reasonably well defined. As the molecules become larger, however, so the necessary approximations become more severe, and the ionization potentials predicted by using Koopmans' theorem may be either higher or lower than the experimental values. For such molecules, the use of semi-empirical SCF methods becomes necessary. The calculations here are simplified further by the neglect of certain integrals and the empirical rather than the theoretical evaluation of others. The calculations involve the use of empirical parameters that are chosen so as to give a good fit either to HF calculations or to some experimental results for a limited number of molecules. The choice of parameters makes it possible to obtain a very good fit in some instances, and the calculations are then expected to have predictive value for other and larger molecules. Semi-empirical calculations are used extensively in the interpretation of photoelectron spectra, although their success in predicting even the ordering of molecular ionizations is varied. The problem for large molecules is certainly very difficult since, because of the high density of electronic states, a relatively small error in an orbital energy can change the order completely. In the simplest theoretical methods of all, Hückel π electron calculations for conjugated molecules, empirical parameters can be chosen specifically to give agreement with photoelectron spectroscopic data. When this is done, the predictive value of the calculations within their restricted range of application is higher than that of any of the other more sophisticated calculations[11].

4.5 COMPARISONS BETWEEN CALCULATION AND EXPERIMENT

4.5.1 HARTREE–FOCK CALCULATIONS

The best calculations of ionization energies possible at present involve finding total HF energies for the neutral molecule and also for the ion in a specified electronic state, and making corrections in both instances for the error in correlation energy. Calculations of this type are available for only one or two molecules, for which, however, excellent agreement with the measured ionization potentials was obtained[12, 13]. There are also very few instances in which the energies of ionic states have been calculated separately, so for all

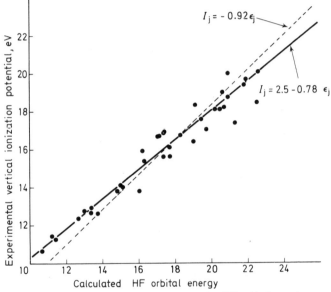

Figure 4.1. Test of Koopmans' theorem for HF orbital energies

practical purposes one must rely on Koopmans' theorem. The purpose of comparing the HF orbital energies with ionization potentials for small molecules is then to establish how well this approximation works out. In *Figure 4.1*, experimental vertical ionization energies for 14 diatomic, triatomic and linear tetra-atomic molecules made up from first row atoms plus chlorine are plotted against HF orbital energies taken from the compendium of Krauss[14].

The correlation is clear but the points are relatively scattered about the estimated best straight line. The spread of the points is around ± 1 eV from the line in the middle and increases for higher ionization energies. The scatter is so large that an exact statistical treatment is not warranted; the data are best represented by a straight line of the form

$$I_{vert.} = 2.5 - 0.8 \, \varepsilon_j \qquad (4.2)$$

(note that the orbital energies ε_j are negative quantities).
The simpler relationship

$$I_{vert.} = -0.92 \, \varepsilon_j \qquad (4.3)$$

which has been used by several workers in comparing *ab initio* SCF calculations with photoelectron spectra[15], does not fit so well, but gives reasonable agreement for ionization potentials between 14 and 18 eV.

From the spread of points from the curve of *Figure 4.1*, it would seem that experimental vertical ionization energies must be separated by about 2 eV, or calculated orbital energies by 2.5 eV, for one to be sure that the theoretical and experimental ordering will agree. Among the 37 ionization energies included in *Figure 4.1*, there are only three instances of disagreement of the ordering within the spectrum of a molecule, but as these are all for small molecules whose ionization bands are on average far apart no severe test is involved. In considering calculations on larger molecules, one must conclude that even if they approach HF accuracy, the ordering of orbitals separated in energy by 1 eV or less is as likely as not to differ from the experimental ordering of ionic states.

4.5.2   *AB INITIO*   SCF CALCULATIONS

This type of calculation can be made for substantially larger molecules than the exact HF calculations[16]; as examples to illustrate the use of *ab initio* calculations in the analysis of photoelectron spectra, the spectra of xenon difluoride and benzene can be considered.

Xenon difluoride

The photoelectron spectrum of XeF_2 has been measured by Brundle *et al.*[17] using excitation by both He I and He II radiation, and the spectra are shown in *Figure 4.2*. The ionic states were identified with the help of orbital energies that were calculated by using a contracted Gaussian basis set in an all-electron SCF method. The calculated orbital energies scaled by the empirical factor of 0.92, together with the measured ionization energies, are shown in *Table 4.1*. Almost all of the experimental vertical ionization energies are

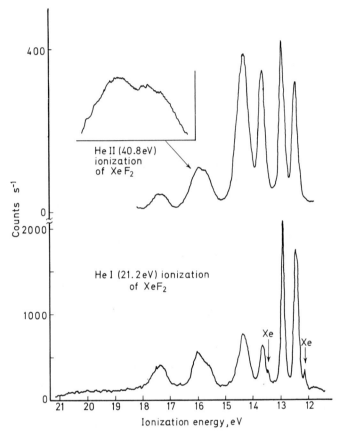

Figure 4.2. Photoelectron spectra of xenon difluoride excited by He I and He II light. (From Brundle *et al.*[17], by courtesy of the American Institute of Physics)

separated from one another by less than 2 eV, so the use of Koopmans' theorem cannot be relied upon for the ordering of the ionic states. An assignment of the spectrum can be made, however, by taking account of the molecular orbital calculations in conjunction with the details of the photoelectron spectrum itself. The highest occupied π orbital of XeF_2 is predicted to be $5\pi_u$, a weakly antibonding orbital with the greatest density at the Xe atom. The $^2\Pi_u$

state resulting from ionization out of this orbital should produce two bands of equal intensity in the spectrum because of the spin–orbit splitting characteristic of the p orbitals of Xe. The expected splitting can be estimated from the calculated Xe atomic population in the $5\pi_u$ orbital together with the characteristic spin–orbit splitting of Xe 6p orbitals (see Chapter 6). The estimated splitting of 0.54 eV

Table 4.1 IONIZATION POTENTIALS AND ORBITAL ENERGIES OF XeF$_2$
(From Brundle et al.[17], by courtesy of the American Institute of Physics)

Orbital	Calculated orbital energies × 0.92, eV	Vertical ionization energy, eV	Ionic state
$5\pi_u$	12.51	12.42	X $^2\Pi_{\frac{3}{2}u}$
		12.89	$^2\Pi_{\frac{1}{2}u}$
$10\sigma_g$	11.79	13.65	A $^2\Sigma_g$
$3\pi_g$	14.71	14.35	B $^2\Pi_g$
$4\pi_u$	15.92	15.60	C $^2\Pi_{\frac{3}{2}u}$
		16.00	$^2\Pi_{\frac{1}{2}u}$
$6\sigma_u$	16.93	17.35	D $^2\Sigma_u$
$9\sigma_g$	25.24	22.5	E $^2\Sigma_g$

agrees sufficiently well with the splitting between the first two peaks in the photoelectron spectrum (0.45 eV) to permit a confident assignment of these two peaks to the $5\pi_u^{-1}$ ionization. Ionization of electrons from the other occupied π molecular orbitals of XeF$_2$, $3\pi_g$ and $4\pi_u$, should give ionic states with spin–orbit splittings of 0 and 0.37, respectively. The band near 16 eV ionization energy is split by about 0.4 eV, in agreement with the prediction for $4\pi_u^{-1}$ ionization, and the broadness of the band is in keeping with the strongly bonding character of this orbital indicated by the calculations. The $3\pi_g^{-1}$ ionization must be intermediate in energy between the two π_u^{-1} ionizations, and must give rise to one of the bands at 13.65 or 14.35 eV. The only remaining orbital whose ionization energy can lie in this region is $10\sigma_g$, which is calculated to have a lower ionization energy. As the band at 14.35 eV is approximately twice as intense as that at 13.65 eV in both the He I and He II spectra, it is assigned to ionization from the doubly degenerate $3\pi_g$ orbital, and the band at 13.65 eV must be the $10\sigma_g^{-1}$ ionization. The remaining bands in the spectrum at 16.93 and 25.24 eV can be attributed to ionization from the $6\sigma_u$ and $9\sigma_g$ orbitals, respectively, on the basis of the calculations, as these are well separated in energy and are the only remaining valence orbitals. The final assignment given in *Table 4.1* is confirmed by the changes in relative band intensities on going from He I (584 Å) to He II (304 Å) excitation. The cross-section for ionization of atomic Xe decreases very considerably between 584 and 304 Å, whereas that for Ne, which can serve as a model for the fluorine atoms, reaches a maximum near the He II wavelength. Hence the ionization from orbitals with the largest Xe atomic contribution would be expected to have a lower relative intensity in the He II spectrum than in that excited by He I. If the π_g^{-1} band is taken as a basis, it can be seen that in fact the $5\pi_u^{-1}$, $4\pi_u^{-1}$ and $6\sigma_u^{-1}$ bands do lose intensity considerably, while the $10\sigma_g^{-1}$ band remains of almost the same intensity. This result is in exact agreement with expectation because on symmetry grounds π_g and σ_g contain no contributions from Xe 5p.

The final assignment of the photoelectron spectrum based on the preceding experimental and theoretical considerations agrees reasonably well with the SCF orbital energy scheme interpreted by Koopmans' theorem, and the theoretical and experimental orderings differ at only one point; nevertheless, it cannot be said that the SCF calculations were essential for the analysis. In fact, the photoelectron spectrum of XeF_2 was measured and analysed independently without the benefit of the SCF calculations by Brehm et al.[18], who reached exactly the same identification of orbitals with ionization bands. The situation with XeF_2 is typical of much work in photoelectron spectroscopy at present in that accurate SCF calculations are an aid to, but by no means essential for, the correct interpretation of experimental spectra.

Benzene

A second example of the comparison between orbital energies calculated by the SCF method and photoelectron spectra is provided by the important and still somewhat controversial case of benzene. The first assignment of the benzene spectrum was made by Jonsson and Lindholm[19], and was based on several different types of information, including SCF calculations. Since then, more experimental evidence has been accumulated, and a wider range of evidence can now be brought to bear on the photoelectron spectrum of benzene than on that of almost any other molecule. Molecular orbital models from the simplest up to ab initio SCF calculations[19, 20], measurements of band intensities and the comparison of spectra excited by He I, He II and Ne I light[21], detailed analyses of the vibrational structure of resolved bands[22, 23], angular distributions of the photoelectrons[24] and comparisons with Rydberg series and energy loss spectra[19, 23] can be used.

The general appearance of the spectrum is shown in *Figure 4.3*; there are seven clearly separate bands below 30 eV ionization energy, a range which covers the whole valence electron ionization region. The number of valence orbitals, however, is 10, so according to the rule that ionization from each orbital gives at least one band in the

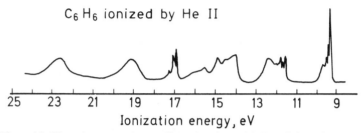

Figure 4.3. Photoelectron spectrum of benzene taken using He II light and corrected for variations of analyser sensitivity with electron energy. There is another broad band near 26 eV which is not shown. (After Åsbrink *et al.*[21], by courtesy of North Holland Publishing Company)

spectrum, several bands must be overlapping. The overlapping bands can be identified both by the total band areas under He II excitation using the rule that the intensity per electron pair is constant, and by the variation of the angular distribution of electrons across the spectrum[24].

From these studies, 10 distinct ionizations can be distinguished, in agreement with the number of valence orbitals. The orbitals are

Table 4.2   VALENCE ORBITALS OF BENZENE
(After B. O. Jonsson and E. Lindholm, *Ark. Fysik*, **39**, 65 (1969))

Orbital	Type	Character
$1\,e_{1g}$	$p\pi$	C—C bonding
$3\,e_{2g}$	$p\sigma$	Weakly C—H and C—C bonding
$1\,a_{2u}$	$p\pi$	Strongly C—C bonding
$3\,e_{1u}$	$p\sigma$	Strongly C—H bonding
$1\,b_{2u}$	$p\sigma$	Strongly C—C bonding
$2\,b_{1u}$	$s\sigma$	Strongly C—H bonding and C—C antibonding
$3\,a_{1g}$	$p\sigma$	Strongly C—H bonding, weakly C—C antibonding
$2\,e_{2g}$	$s\sigma$	Weakly C—H bonding and C—C antibonding
$2\,e_{1u}$	$s\sigma$	Strongly C—C bonding
$2\,a_{1g}$	$s\sigma$	Strongly C—C bonding

named and described in *Table 4.2* and the experimental ionization energies are given in the first column of *Table 4.3*. The measured band intensities also show where degenerate orbital ionizations are involved, and this information is given in the second column of *Table 4.3*.

The molecular orbital calculations can be used to identify some of the bands directly. The first ionization at 9.3 eV is identified with loss of the outermost π electron, $1e_{1g}^{-1}$, in all calculations; there is also experimental evidence for this from a study of the corresponding Rydberg series. Empirical π electron-only calculations agree with this assignment, and indicate that the second π ionization, $1a_{2u}^{-1}$, must occur at about 12 eV, and therefore corresponds to one component of the second spectral band. Next, the ionizations from orbitals based mainly on carbon 2s atomic orbitals can be recognized as they lie at higher ionization energies than those based on carbon 2p. Both the SCF calculations and empirical molecular orbital arguments indicate that the bands at 26, 22.5 and 19.5 eV correspond to $2a_{1g}^{-1}$, $2e_{1g}^{-1}$ and $2e_{2g}^{-1}$ ionizations, respectively, in agreement with the degeneracies from experimental band intensities. The one remaining carbon 2s-based orbital, $2b_{1u}$, is calculated to lie near 16 eV, and must be one of the ionizations at 16.9 or 15.4 eV. These conclusions from the molecular orbital models are shown in the third column of *Table 4.3*.

Four bands in the photoelectron spectrum have resolved vibrational structure, and all have been examined at high resolution both in the spectrum of benzene and in that of hexadeuterobenzene. The vibrational structures of the bands at 15.4 and 16.9 eV contain progressions in both the C–H and C–C stretching vibrations, which shows that the orbitals concerned have C–H bonding character. The only non-degenerate orbitals with strong C–H bonding character are $2b_{1u}$, which has already been assigned to one of these two ionization bands, and $3a_{1g}$, so the two bands comprise ionization from these two orbitals; further evidence must be used to distinguish between the two bands. Firstly, the Rydberg defect of the series converging on the 16.9 eV ionization is 0.45, which is too large for d-type Rydberg orbitals, the only orbitals allowed by the selection rules if a $2b_{1u}$ s-type orbital is ionized. This shows that the 16.9 eV band must be the $3a_{1g}^{-1}$ ionization, for which the Rydberg orbitals are of p-type. Confirmation is provided by comparison with the spectrum of furan, which contains a band at 17 eV very similar to the 16.9 eV band of benzene, and for which only a p atomic orbital origin is possible. On these grounds, the band at 16.9 eV is assigned to ionization from $3a_{1g}$ and that at 15.4 eV to ionization from $2b_{1u}$.

The vibrational structure of the first band in the benzene spectrum at 9.3 eV contains progressions in a degenerate vibrational mode, v_{18}; the excitation of such a degenerate vibration is the first sign of a Jahn–Teller effect (see Chapter 6), and is further proof that the orbital involved is degenerate. The second band also has vibrational

Table 4.3 ANALYSIS OF THE BENZENE SPECTRUM

Vertical ionization energy, eV	Band intensity	MO calculations	Deductions from			Final assignment
			Vibrational structure	Rydberg defect	Angular distribution	
9.3	Degen.	e_{1g}	Degen.	$1\,e_{1g}$		$1\,e_{1g}$
11.4	One degen.	One $1\,a_{2u}$	Degen.?			$3\,e_{2g}$
12.1	One single					$1\,a_{2u}$
13.8	One degen.					$3\,e_{1u}$
14.7	One single				$3\,e_{1u}$	$1\,b_{2u}$
15.4	Single	One $2\,b_{1u}$	$\left\{\begin{array}{l}3\,a_{1g}\\2\,b_{1u}\end{array}\right.$			$2\,b_{1u}$
16.9	Single			$3\,a_{1g}$		$3\,a_{1g}$
19.2	Degen.	$2\,e_{2g}$				$2\,e_{2g}$
22.5	Degen.	$2\,e_{1u}$				$2\,e_{1u}$
25.9		$2\,a_{1g}$				$2\,a_{1g}$

structure similar to that of the first band, and Åsbrink *et al.*[22] attribute the vibration of frequency 640 cm^{-1} to the same degenerate vibration, v_{18}. This would be proof that the orbital concerned is degenerate and cannot be $1a_{2u}$. There are also Rydberg series that converge on the first and second ionization potentials, both with quantum defects of 0.46, very close to the defect in the series that converge on $3a_{1g}^{-1}$. The Rydberg orbitals for $3a_{1g}^{-1}$ and $1e_{1g}^{-1}$ ionizations are both p-type by symmetry, and the fact that no p-type

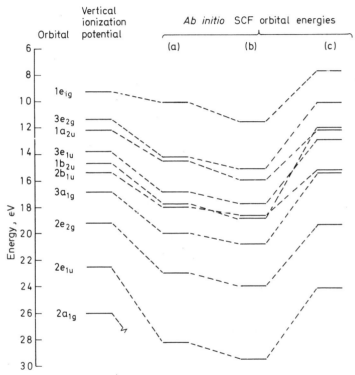

Figure 4.4. The ionization potentials of benzene compared with orbital energies from *ab initio* SCF calculations. The calculations are from (a) Schulman, J. M. and Moskowitz, J. W., *J. chem. Phys.*, **47**, 3491 (1967); (b) Prand, L., Millie, P. and Berthier, G., *Theor. Chim. Acta*, **11**, 169 (1968); and (c) Schulman, J. M. and Moskowitz, J. W., *J. chem. Phys.*, **43**, 3287 (1965)

Rydberg series is allowed in $1a_{2u}^{-1}$ ionization again points to the conclusion that the second ionization cannot be $1a_{2u}^{-1}$. This conclusion is not final, however, as other evidence has been found that points to the identification of the band at 11.4 eV with a_{2u}^{-1} ionization[23]. The second band must in any case include a_{2u}^{-1} and one ionization from a degenerate orbital.

The ionizations that remain unassigned are one component of the overlapping second band, 11.4 or 12.1 eV, and both components of the overlapping third band, at 13.8 and 14.7 eV. In both bands, one component must be degenerate and one non-degenerate according to the band areas. The analysis follows from the angular distribution measurements of Carlson and Anderson[24], because, of the three remaining orbitals, $3e_{1u}$, $3e_{2g}$ and $1b_{2u}$, only one, $3e_{1u}$, is predicted to give an angular distribution characterized by a high β value on ionization. Only the band at 13.8 eV has a high β value experimentally, and this band must be the $3e_{1u}^{-1}$ ionization; the degenerate nature of this band is confirmed by its Jahn–Teller contour, as there are two maxima within the range of the high β value. Once $3e_{1u}^{-1}$ has been assigned, the 14.7 eV ionization must be $1b_{2u}^{-1}$ and the degenerate component of the first band must be $3e_{2g}^{-1}$. The analysis is now complete, and is shown in the last column of *Table 4.3*.

The experimental assignment is compared in *Figure 4.4* with orbital energies obtained in three *ab initio* SCF calculations, which differ mainly in the forms of the basis sets. The orderings of the SCF orbital energies agree reasonably well with the order deduced from the photoelectron spectrum using Koopmans' theorem, and only in the region where the density of electronic states is highest is there any significant discrepancy. The two most recent calculations, (a) and (b), even reproduce the pattern of spacings between the orbitals relatively well, but nevertheless the SCF orbital energies played only a minor role in the analysis of the spectrum. The characters of the different orbitals, on the other hand, which were obtained from the wave-functions, were essential for the assignment. It seems to be generally true that theoretical predictions of orbital character are more reliable than calculations of orbital energy.

4.5.3 SEMI-EMPIRICAL CALCULATIONS

Semi-empirical calculations involve further simplifications to the HF method, and because of the simplifications achieved they can be applied to large and complex molecules. The different methods that have been extensively used in comparisons with photoelectron spectra are generally known by their acronyms, CNDO, INDO, MINDO and SPINDO, where NDO = neglect of differential overlap, C = complete, I = intermediate, M = modified and SP = spectroscopic potentials adjusted. The semi-empirical nature of these calculations consists in the replacement of certain integrals by parameters that are chosen so that the theory agrees with experiment (or with better SCF calculations) for certain properties of a restricted range of standard molecules. It is then hoped that the calculations will have a high predictive value for these and other properties of a whole range of molecules for which they can be carried out, a hope which is sometimes fulfilled. As far as calculations of ionization energies from orbital energies by Koopmans' theorem are concerned, however, they are at best not superior to the *ab initio* SCF calculations. Like the SCF calculations, they can be very useful in indicating the characters of different orbitals and hence the nature of the corresponding bands to be expected in the spectrum. Band widths, spin–orbit splittings or Jahn–Teller splittings can often be estimated more reliably than orbital energies.

The SPINDO method is a new procedure[25], and is the first semi-empirical calculation to be parameterized directly on ionization potentials from photoelectron spectroscopy. There is some reason to hope that its predictive value for ionization energies might be higher than that of the other methods, but it is too recent to have been properly evaluated. The comparison of photoelectron spectra with semi-empirical calculations involves no new principles, so no further examples are given. A review by Worley[16] contains and refers to a large range of examples up to 1970, but excludes SPINDO, which has so far been used in only a few instances[26].

4.5.4   EMPIRICAL CALCULATIONS

In empirical molecular orbital calculations, no attempt is made to mimic the HF method; the aim is rather to evolve the simplest possible theory, which can nevertheless be used to correlate a wide range of experimental results while using only a small number of arbitrary parameters. The best empirical method is Hückel molecular orbital (HMO) theory[2], which is applied to the π electron systems of conjugated and aromatic molecules, and has just two parameters. It has a most impressive record of success in correlating a wide range of physical and chemical properties of such compounds, both one-electron properties, such as the forms of the ultraviolet spectra and electron spin resonance spectra, and also properties that depend on the total π electron energy such as thermodynamic constants and many chemical reaction rates. Its success in interpreting photo-electron spectra of π electron molecules via Koopmans' theorem is no less impressive.

The simplest hydrocarbon molecules that contain π electrons are the acetylenes, in which the σ and π electron systems are completely separable because of the linear structure of the molecules. The photoelectron spectra of acetylene and diacetylene[27] are shown in *Figure 4.5.*

All models are in agreement that the outermost electrons in acetylene are in the degenerate π orbital, so the band with ionization potential at 11.8 eV in acetylene can confidently be attributed to the π_u^{-1} ionization. The vibrational structure excited in this ionization consists of a simple progression in 1830 cm^{-1}, which is identified as

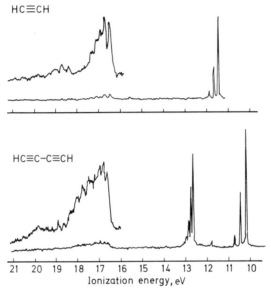

Figure 4.5. Photoelectron spectra of acetylene and diacetylene. (After Baker and Turner[27], by courtesy of the Chemical Society)

the vibration v_2. This is the symmetrical stretching of the C–C bond, exactly as expected if the effect of removing a π electron is simply to weaken this bond so that it is longer in the ion than in the molecule. When two acetylene molecules are joined together in diacetylene, two occupied π orbitals appear, one of them lower and one higher in energy than the π orbital of acetylene. The two bands in diacetylene that are placed symmetrically about the ionization potential of acetylene can confidently be attributed to ionization from these two new orbitals. The symmetrical placing of two π^{-1} ionization bands in the spectrum of diacetylene about the ionization potential of acetylene is not predicted by the simplest HMO theory, which completely omits overlap integrals[2] and treats all bonds, whatever their length, as equivalent. If either of these two approximations is dropped, the observed pattern can be reproduced. The patterns of π molecular orbital energies for acetylene and diacetylene are, however, so simple that they could be predicted without any calculation from the empirical idea of how two new orbitals are created when two systems interact. This is no longer so for the larger conjugated hydrocarbons and here HMO calculations are invaluable in interpreting the photoelectron spectra. Aromatic hydrocarbons from naphthalene[28] to ovalene[29] have now been studied and their π electron ionization bands have been identified and assigned to definite ionic states by comparison with HM calculations. The

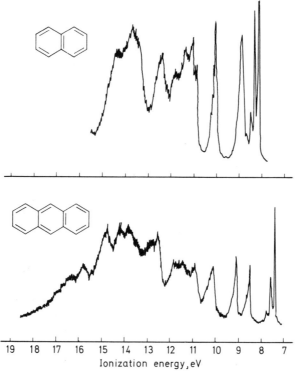

Figure 4.6. Photoelectron spectra of naphthalene and anthracene. (From Clark, Brogli and Heilbronner[36], by courtesy of the Swiss Chemical Society)

spectra of naphthalene and anthracene are shown in *Figure 4.6* as examples for this class of compound. Typical features are the sharp bands at low ionization potential, all of which correspond to π electron ionizations, followed by a broad and unresolved structure at energies greater than 10 eV, corresponding to ionization from the σ orbitals. The ionization energy for the highest occupied σ orbital does not vary much from naphthalene, where it is 11 eV, to the largest molecules studied, where continuous bands begin at about 10.5 eV. The first ionization energy decreases from 9.25 eV in benzene to 8.12 eV in naphthalene and 6.74 eV in pentacene, so the outermost orbitals are predominantly of the π type. The semi-empirical calculations (particularly INDO and its progeny) do not represent this result at all well, as they generally locate one or more σ orbitals in the middle of the π ionization region[16].

In HMO theory, the orbital energy ε_j of the jth π orbital is given by an expression

$$\varepsilon_j = \alpha + m_j \beta \qquad (4.4)$$

where m_j is the Hückel coefficient, which can easily be calculated or obtained from standard tables[30], and α and β are the only two arbitrary parameters, called the coulomb and resonance integrals. Brogli and Heilbronner[11] have compared 34 experimental ionization potentials for unsaturated and conjugated compounds ranging from ethylene to phenanthrene with the HMO orbital energies, and found the linear regression

$$I_j = 6.553 \pm 0.34 + m_j (2.734 \pm 0.333) \text{ eV} \qquad (4.5)$$

The uncertainties correspond to 90% confidence limits. Only for ethylene does the experimental ionization energy differ by more than 1 eV from the calculated value, a result which compares favourably with that obtained in HF calculations on small molecules.

The agreement between theory and experiment reached by the simplest HMO theory is already good, and enables most π electron ionization bands to be identified with reasonable certainty. Nevertheless, the deviations from exact agreement are interesting. In anthracene, it is very noticeable that whereas HMO theory predicts two pairs of degenerate orbitals, the photoelectron spectrum (*Figure 4.6*) contains only single bands (from their intensities) in the region where ionization from one of the degenerate pairs is expected. The predicted degeneracy is not a result of molecular symmetry but of the simplifying assumptions of HMO theory, particularly that all bonds are exactly equivalent. That this is not in fact true is easiest to visualize by using the valence bond model of molecular structure, because when the canonical forms for anthracene are drawn, double bonds do not occur in all possible positions with equal frequency. Compensation for the effect of this bond fixation on the ionization potentials can be made by a first-order perturbation treatment[11], which yields the equation

$$I_j = a + m_j b_1 + y_j b_2 \qquad (4.6)$$

where m_j has the same meaning as in equation 4.4, y_j is a similar calculated coefficient and a, b_1 and b_2 are three empirical parameters. The fit of the measured ionization potentials to this modified HMO theory is excellent, the maximum deviation between theory and calculation for any of the sample ionization potentials being less than 0.5 eV. None of the SCF methods used on the same molecules and interpreted by Koopmans' theorem gives such a close fit, and even re-minimization of the ionic states in order to calculate first ionization potentials is not significantly better. The modified HMO model is an excellent example of an empirical theory of high predictive value for ionization potentials, and has been used in the interpretation of several photoelectron spectra.

The basic Hückel model discussed above is applicable only to π electrons, but it is also possible to use a Hückel type of formalism for all electrons. One theory of this type is the extended Hückel (EH) theory[31], which has been used to some extent in the interpretation of photoelectron spectra, although not with great success[16]. A simpler Hückel formulation for all electrons has recently been used by Cox et al.[32], who showed that it has several advantages over more sophisticated methods. For the interpretation of photoelectron spectra, it is possible that use of a simple empirical theory may be the best approach.

4.6 IONIZATION POTENTIALS AND MOLECULAR CHARGE DISTRIBUTIONS

In X-ray photoelectron spectroscopy, the ionization potentials of inner shell electrons are measured and are found to vary slightly with the chemical environment of the atoms in a molecule[33]. Because there are essentially no direct interactions between orbitals of the inner shells, these chemical shifts are due to changes in the electrical potential of the inner electrons caused by charges on the atoms within the molecules. Extensive comparisons of observed shifts with calculated atomic charge distributions have shown that the relationship between the shift and charge is almost linear. This result agrees with an electrostatic model[34], which indicates that a charge q on a sphere of radius r produces a change of potential $\Delta U = q/r$ throughout the interior of the sphere. The chemical shift on a particular atom should be the same for all inner shells, and this is also found to be true. In a real molecule, the charge removed from one atom must finally be located on another or on several other atoms, so that the potential must include contributions from these more distant charges, a Madelung potential. The change in potential can be written as

$$\Delta U = \frac{q}{r} + \sum_i \frac{\delta q_i}{R_i} \qquad (4.7)$$

where R_i is the distance from the atom of interest to the ith other atom carrying charge δq_i. If the ionization potential shift, ΔU, is measured, the atomic charge, q, can be calculated provided that the constant $1/r$ is known and the Madelung potential can be estimated.

The relevance of these remarks to ultraviolet photoelectron spectroscopy is that even with 21.2 eV radiation some inner shell ionizations can be seen, and with 40.8 eV light the range will be

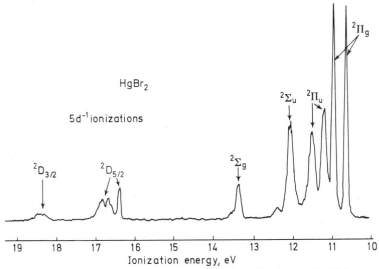

Figure 4.7. Photoelectron spectrum of mercury(II) bromide, showing the 5d electron ionizations. (From Eland[35], by courtesy of Elsevier Publishing Company)

greatly extended. This means that ultraviolet photoelectron spectroscopy offers the possibility of deducing the charge distribution in several molecules and thence also the partial ionic characters of the bonds.

The inner shell ionizations that one can expect to observe are those of the full d shells of the B-group metals (Cu, Ag, Au, Zn, Cd and Hg with He I), those of the full p shells of the alkali metal and alkaline earth metal atoms (K and Ba with He I) and the full f shells of the third-row transition metals. The reason why few inner shell ionizations have been observed so far is simply that most compounds of these elements are too involatile to be introduced into unmodified photoelectron spectrometers. Only divalent mercury compounds[35] and some zinc and cadmium compounds have been investigated from this point of view up to now. As examples of the photoelectron spectra, those of mercury(II) bromide and dimethylmercury are shown in *Figure 4.7* and *4.8.* The d electron ionizations

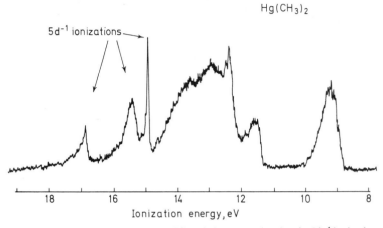

Figure 4.8. Photoelectron spectrum of dimethylmercury, showing the $5d^{-1}$ ionization bands above 15 eV ionization energy. (From Eland[35], by courtesy of Elsevier Publishing Company)

produce three bands in the spectra at ionization potentials above 15 eV; for mercury(II) bromide they are 16.4, 16.8 and 18.3 eV and for dimethylmercury 15.0, 15.4 and 16.9 eV. These bands are attributed to Hg $5d^{-1}$ ionizations on the following grounds:
  (1) they appear in the spectrum of every mercury compound examined;
  (2) the relative spacing of the three bands remains almost constant, although the absolute position on the energy scale varies;
  (3) the spacing of the two outer bands is close to 1.86 eV, the spin–orbit splitting of the ^2D state of Hg^+.

Of the three bands, the one at the lowest ionization potential is always sharp, and it represents ionization from a d orbital perpendicular to the axis of the molecule. None of the valence orbitals of the molecule can interact with this orbital, so it is essentially nonbonding. The ionic state produced has a total electronic angular momentum of $\pm\frac{5}{2}$; it corresponds to one component of the $^2D_{\frac{5}{2}}$ state of the atomic mercury ion. The second band is clearly split in the spectra of some compounds and corresponds to the $J = \frac{3}{2}$ and

$J = \frac{1}{2}$ components of $^2D_{\frac{5}{2}}$, which do experience some interactions with the valence electrons in the molecule. The third band, which is also sometimes split, corresponds similarly to the formation of $J = \frac{3}{2}$ and $J = \frac{1}{2}$ states derived from $^2D_{\frac{3}{2}}$ of Hg$^+$. The classification of electronic states according to J value rather than according to orbital angular momentum is necessary because the splitting of the ^2D state by spin–orbit interactions is much larger than that caused by the valence interactions.

The shift between the sharp $J = \frac{5}{2}$ peak and the $^2D_{\frac{5}{2}}$ ionization potential of atomic mercury is a direct measure of the difference between the mercury atom in the molecule and a free mercury atom. The shifts can be equated with changes in potential and used in equation 4.7 to deduce the charge on the mercury atom, once the constant $1/r$ is known. This constant is equal to the change in ionization potential of the d shell produced by unit positive charge in the outer 6s shell, and it can be obtained from the atomic spectrum of mercury[35]. Some atomic charges deduced in this way for mercury(II) halides and other mercury(II) compounds are given in *Table 4.4*. The atomic charges are closely related to such parameters

Table 4.4   ATOMIC CHARGES FROM 5d^{-1} IONIZATION POTENTIALS IN MERCURY(II) COMPOUNDS

Compound	Shift from $^2D_{\frac{5}{2}}$, eV	Charge on Hg (units of e)
HgCl$_2$	1.87	0.425
HgBr$_2$	1.56	0.332
HgI$_2$	1.15	0.230
MeHgCl	0.95	0.218
MeHgBr	0.84	0.182
MeHgI	0.64	0.134
HgMe$_2$	0.11	0.028
HgEt$_2$	−0.16	−0.041

as the partial ionic characters of the bonds and the electronegativities of the different atoms and groups, which can readily be derived from the measurements. The change in sign of the charge on mercury between the dimethyl and diethyl compound shows that the electron-attracting power of a mercury atom is intermediate between that of the methyl and the ethyl groups. The availability of calibrations of equation 4.7 from atomic spectra is an important general advantage of ultraviolet over X-ray photoelectron spectroscopy in interpreting chemical shifts of ionization potentials. It means that the atomic charges deduced are independent of theory, apart from the electrostatic model itself, whereas in X-ray work the calibration is usually made against charges calculated according to theoretical models of molecular electronic structure. If there is a charge of $+0.425$ units on the mercury atom in mercury(II) chloride, the chlorine atoms must each have a charge of -0.2125 units, and this ought to be reflected in the chlorine electron ionization potentials also. No chlorine inner shell electrons can be ionized with He I light, and so it is interesting to inquire if the charge effect can be deduced from the ionization potentials of non-bonding chlorine electrons of the valence shell. This is not yet possible quantitatively, although the existence of shifts in valence shell ionizations caused by the presence of charges is clear. The chlorine lone-pair ionization potential in HCl is at 12.4 eV, in $HgCl_2$ at 11.4 eV and in methyl chloride it decreases to 11.22 eV. The difficulty is to separate the shifts due to charges from shifts caused by conjugative and other interactions, and this separation cannot be achieved experimentally.

REFERENCES

1. BREHM, B., Nineteenth Annual Conference on Mass Spectrometry and Allied Topics, Atlanta, Georgia, American Society for Mass Spectrometry, 78 (1971)
2. MURRELL, J. N., KETTLE, S. F. A. and TEDDER, J. M., *Valence Theory*, John Wiley, London (1965)
3. POPLE, J. A., *Q. Revs.*, 11, 273 (1957)
4. KOOPMANS, T., *Physica*, 1, 104 (1933)
5. NEWTON, M. D., *J. chem. Phys.*, 48, 2825 (1968)
6. RICHARDS, W. G., *Int. J. Mass Spectrom. Ion Phys.*, 2, 419 (1969)
7. DODDS, J. L. and MCWEENEY, R., *Chem. Phys. Lett.*, 13, 9 (1972)
8. CRADOCK, S. and WHITEFORD, R. A., *J. chem. Soc., Faraday Trans. II*, 68, 249 (1972)
9. COX, P. A. and ORCHARD, A. F., *Chem. Phys. Lett.*, 7, 273 (1970)
10. CLEMENTI, E., *Chem. Rev.*, 68, 341 (1968)
11. BROGLI, F. and HEILBRONNER, E., *Theor. Chim. Acta*, 26, 289 (1972)
12. RICHARDS, W. G. and WILSON, R. C., *Trans. Faraday Soc.*, 64, 1729 (1968)
13. VERHAEGEN, G., RICHARDS, W. G. and MOSER, C. M., *J. chem. Phys.*, 47, 2595 (1967)
14. KRAUSS, M., *N.B.S. Tech. Note No. 438*, U.S. Government Printing Office, Washington, D.C. (1967)
15. BRUNDLE, C. R., ROBIN, M. B., BASCH, H., PINSKY, M. and BOND, A., *J. Amer. chem. Soc.*, 92, 3863 (1970)
16. WORLEY, S. D., *Chem. Rev.*, 71, 295 (1971)
17. BRUNDLE, C. R., ROBIN, M. B. and JONES, G. R., *J. chem. Phys.*, 52, 3383 (1970)
18. BREHM, B., MENZINGER, M. and ZORN, C., *Can. J. Chem.*, 48, 3193 (1970)
19. JONSSON, B. Ö. and LINDHOLM, E., *Ark. Fysik*, 39, 65 (1969)
20. PRAND, L., MILLIE, P. and BERTHIER, G., *Theor. chim. Acta*, 11, 169 (1968)
21. ÅSBRINK, L., EDQVIST, O., LINDHOLM, E. and SELIN, L. E., *Chem. Phys. Lett.*, 5, 192 (1970)

22. ÅSBRINK, L., LINDHOLM, E. and EDQVIST, O., *Chem. Phys. Lett.*, **5**, 609 (1970)
23. POTTS. A. W.. PRICE. W. C., STREETS. D. G. and WILLIAMS. T. A.. *Discuss. Faraday Soc.*, **54**, 168 (1973)
24. CARLSON, T. A. and ANDERSON, C. P., *Chem. Phys. Let.*, **10**, 561 (1971)
25. LINDHOLM, E., FRIDH, C. and ÅSBRINK, L., *Discuss. Faraday Soc.*, **54**, 127 (1973)
26. FRIDH, C., ÅSBRINK, L. and LINDHOLM, E., *Chem. Phys. Lett.*, **15**, 408 and 567 (1972)
27. BAKER, C. and TURNER, D. W., *Chem. Commun.*, 797 (1967)
28. ELAND. J. H. D. and DANBY, C. J.. *Z. Naturforsch.*, **23a**, 355 (1968)
29. BOSCHI, R., MURRELL, J. N. and SCHMIDT, W., *Discuss. Faraday Soc.*, **54**, 116 (1973)
30. COULSON, C. A. and STREITWEISER, A., JR., *Dictionary of π-Electron Calculations*, Pergamon Press, London (1965)
31. HOFFMANN, R., *J. chem. Phys.*, **39**, 1397 (1963)
32. COX, P. A., EVANS, S., ORCHARD, A. F., RICHARDSON, N. V. and ROBERTS, P. J., *Discuss. Faraday Soc.*, **54**, 26 (1973)
33. SIEGBAHN, K., NORDLING, C., FAHLMAN, A., NORDBERG, R., HAMRINN, K., HEDMAN, J., KOHANSSON, G., BERGMARK, T., KARLSSON, S. E., LINDGREN, I. and LINDBERG, B., *Nova Acta Regiae Soc. Sci. Upsal., Ser. IV*, **20** (1967)
34. FAHLMAN. A.. HAMRIN, K., HEDMAN, K., NORDBERG, J., NORDLING. R. and SIEGBAHN. K.. *Nature. Lond.*, **210**, 4 (1966)
35. ELAND, J. H. D., *Int. J. Mass Spectrom. Ion Phys.*, **4**, 37 (1970)
36. CLARK, P. A., BROGLI, F. and HEILBRONNER, E., *Helv. Chim. Acta*, **55**, 1415 (1972)

5 Photoelectron Band Structure—I

5.1 INTRODUCTION

The position of a photoelectron band indicates the energy of an ionic state, whereas the structure of the band gives information about the structure and bonding of the ions in that state. If a band has resolved vibrational fine structure, the identities and frequencies of the modes excited on ionization, and the vibrational line intensities, can be examined. An unresolved band gives less but still useful information in its width and shape. The topics covered in this and the following chapter are the analysis of photoelectron band structures and the deductions that can be made from them. The simpler aspects are covered in this chapter and Chapter 6 concentrates on the complexities of band structure that appear when orbitally degenerate states are reached in ionization.

Because photoionization occurs from the vibrationless ground states of molecules, photoelectron spectra are characteristic primarily of the ionic states produced. Deductions about structure and bonding in the molecules depend on the reliable assumption of the Franck–Condon principle and on the less reliable assumption of Koopmans' theorem. Another matter which is also involved is the question of charge delocalization following ionization. When inner-shell electrons are ionized by X-ray photons, it seems that the charges do not become delocalized over all the equivalent atoms in a molecule on the time scale of the experiment. If a 1s electron is removed from carbon in acetylene, for instance, the kinetic energy of the ejected electron can be calculated accurately only if it is assumed that one carbon atom in the $C_2H_2^+$ ion has a vacancy in

110

its 1s orbital and the other is not affected. When symmetrical charges of $+\frac{1}{2}$ on each carbon atom are assumed, the calculated ionization energy changes by about 6 eV, which is more than the experimental or computational error[1]. On the other hand, ionization of any valence electron from the C_2H_2 molecule certainly results in a symmetrical charge distribution in the ions on the time scale of ultraviolet photoelectron spectroscopy. It is perhaps possible that a changeover occurs as one chooses deeper and deeper lying electrons. Similarly, it is not clear what happens when one lone-pair electron is ionized from a large molecule that contains two equivalent hetero-atoms well separated in space. Localization or delocalization must affect the structure of the molecular ions and hence may also have an observable effect on band structures in photoelectron spectra. These problems have not yet been solved but are now beginning to receive attention.

In discussing the structure of photoelectron bands, it is necessary to refer repeatedly to vibrational modes or normal modes of vibration. The definition of normal modes and their forms in different molecules cannot be included here and must be studied in textbooks on symmetry or infrared and Raman spectroscopy[2].

5.2   ANALYSIS OF VIBRATIONAL STRUCTURE

In order to identify the vibrational modes that are excited on ionization one first compares the vibrational intervals observed in a photoelectron band with the ground state vibrational frequencies of the molecule, which are known from infrared or Raman spectroscopy. It is very rare for the frequencies of the same mode in the molecule and ion to differ by more than a factor of two, and much smaller changes are usual. This alone is not sufficient for positive identifications except in very small or highly symmetrical molecules that have few vibrational modes. A useful adjunct is the fact that both the change in frequency and the change in equilibrium bond lengths following ionization are related to the bonding character of the electron removed. These two effects are also, therefore, related to one another, and if a proposed assignment involves a large change in frequency between molecule and ion, excitation of a long progression in that frequency is to be expected on ionization. This idea has been placed on a semi-quantitative basis by Turner[3], who showed that the fractional change in frequency and the difference between adiabatic and vertical ionization potentials are roughly linearly related. A diagram showing the relationship is reproduced

in *Figure 5.1* and it can be used to check the suitability of possible
identifications.

An important aid to the unravelling of vibrational structure is the
existence of selection rules for vibrational excitation in all electronic
transitions, including ionization. The selection rules indicate those
vibrational modes which should be excited strongly and those excitations which are forbidden in the ionization of molecules with elements of symmetry. These selection rules are based on the Franck–
Condon principle and can be explained as follows. If the electronic

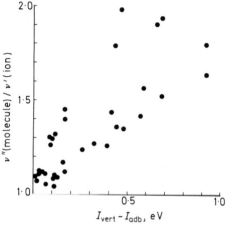

Figure 5.1. Relationship between the change in frequency on ionization and the
vibrational excitation energy. (From Turner[3], by courtesy of the Council of the Royal
Society)

and vibrational motions can be separated, then in an electronically
allowed band* the relative intensity of a vibrational transition from
the vibrational level v'' in the molecule to v' in the ion is given by the
Franck–Condon factor (FCF), which is equal to the square of the
overlap integral between the vibrational wave-functions $\chi_{v''}$ and
$\chi_{v'}$:

$$\text{FCF} = |<\chi_{v'}|\chi_{v''}>|^2 \qquad (5.1)$$

Within one electronic state, the vibrational wave-functions for levels
with different vibrational quantum numbers, v, are orthogonal to
one another, that is, the overlap integrals between them are zero.
If another electronic state has a potential energy surface of the same
shape, the vibrational wave-functions will also be the same, and the

---

* Almost all photoionization bands are electronically allowed in this sense, as
the ionizations that are electronically forbidden, for example two-electron processes,
cannot be made allowed by vibronic interactions.

overlap integrals will be zero for transitions that involve any change in vibrational quantum number. If there is little change in the potential energy surface on ionization, in other words little change in bond lengths or force constants, the only strong transitions are those in which the vibrational quantum number remains the same. As in photoelectron spectroscopy the target molecules are in their vibrational ground states, this means that adiabatic (0–0) transitions are the most intense. If, on the other hand, there is a change in the shape of the potential energy surface on ionization so that the equilibrium positions of the nuclei are different, then the vibrational

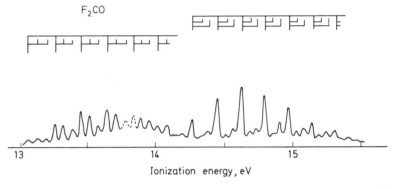

Figure 5.2. Partial photoelectron spectrum of carbonyl fluoride showing progressions of progressions. The vibrational analysis is indicated above. (From Thomas, R. K. and Thompson, H., *Proc. R. Soc., Lond.*, **A327**, 13 (1972), by courtesy of the Council of the Royal Society)

mode or modes that correspond most closely to the change in nuclear positions are most strongly excited. This is the basis of the relationship between the localization of bonding character of particular electrons and the structure of ionization bands.

When two or more vibrational modes are strongly excited in a transition, the band structure consists of progressions of progressions, which means that each line corresponding to the excitation of a number of quanta of one mode is the starting-point for a progression in the other mode. The bands can become very complicated, and the resolution in photoelectron spectroscopy has up to now been such that only if one mode is much less strongly excited than the other have bands been fully assigned. Examples of progressions of progressions are seen in the photoelectron spectrum of carbonyl fluoride, shown in *Figure 5.2*.

5.2.1   VIBRATIONAL SELECTION RULES

All that has been said up to now applies equally to symmetrical and unsymmetrical molecules. For molecules with some elements of symmetry there are more exact selection rules that result in a simplification of the vibrational structure.

If the overlap integral in equation 5.1 is to be non-zero, the product of the vibrational wave-functions must be fully symmetric. In the language of group theory, the direct product of the representations to which $\chi_{v'}$ and $\chi_{v''}$ belong must be or contain the totally symmetric representation in the point group of the molecule and ion; it can only do so for combinations of vibrational species of identical symmetry. In ionization from the vibrationless ground state of the molecule, which is always totally symmetric, only totally symmetric vibrational levels can be reached. The vibrational wave-functions for symmetric vibrational modes are fully symmetric whatever the vibrational quantum number, so excitation of such modes is allowed with any number of quanta. In excitation of anti-symmetric vibrational modes, however, the wave-functions are antisymmetric for odd and symmetric for even vibrational quantum numbers. Such modes can be excited on ionization only in units of two quanta:

$$\Delta v_k = 0, \pm 2, \pm 4 \ldots \tag{5.2}$$

The intensities of the transitions with double quantal excitation are very much less than that of the (0–0) transition, because the vibrational wave-functions in the upper state always have their maxima at the symmetrical position directly above the vibrationless ground state on a Franck–Condon diagram. Even for a change in frequency of an antisymmetric mode by a factor of two, 95% of the total intensity remains in the (0–0) transition. Double quantal excitation of antisymmetric vibrational modes on ionization with the expected low intensity has been observed[4, 5] in the photoelectron spectra of the triatomic molecules CO_2 and CS_2.

Excitation of degenerate vibrational modes on ionization is governed by the same rules as those for the excitation of anti-symmetric modes, and only weak transitions in units of two quanta are allowed. For certain point groups, the odd quantal states of degenerate vibrations are also antisymmetric with respect to at least one element of symmetry and rule 5.2 applies strictly. In other instances, transitions with $\Delta v_k = \pm 3, \pm 5$ may be very weakly allowed, but $\Delta v_k = \pm 1$ is always strictly forbidden. A possible example of double quantal excitation of such a degenerate vibration has been found in the first band in the photoelectron spectrum of ethylene[6].

If ionization causes a change in equilibrium symmetry between molecule and ion, such as bent to linear, non-planar to planar or vice versa, the preceding rules still apply, but only in respect of symmetry elements that are common to the initial and final state. Hence in a linear to bent transition of a symmetric triatomic molecule, excitation of the bending mode, which is degenerate in the linear form, becomes allowed, but excitation of the antisymmetric stretching mode remains forbidden, except weakly in double quanta.

When ionization produces an orbitally degenerate state of the molecular ion, as it does if ionization is from a degenerate orbital, there is frequently a change in molecular geometry and the vibrational structure may be complex. In transitions to degenerate electronic states, the selection rules previously given apply only if coupling between electronic and vibrational motions in the degenerate state is negligible. Molecules that have orbitally degenerate electronic states also possess degenerate vibrational modes, and interactions that involve the two degeneracies cause complications in the spectra. The complications, when they occur, are severe and comprise the Jahn–Teller effects in non-linear molecules and the Renner–Teller and related effects in linear molecules; these topics are considered in Chapter 6. Coupling of the vibrational and electronic motions means that resolution of the transition moment into electronic and vibrational parts is no longer valid, and the selection rules must be based on the vanishing or non-vanishing of the dipole matrix element $< \Psi \,|\, M \,|\, \Psi >$. The wave-functions Ψ now describe vibronic states, and the electron continuum is included in the final state wave-function for a photoionization process.

In the photoelectron spectra of linear molecules, there are many examples of ionization to degenerate electronic states where no complications are seen and the normal selection rules are obeyed. In the spectra of non-linear molecules, on the other hand, bands due to degenerate electronic states usually do involve complications that can be attributed to Jahn–Teller effects.

5.3   INTERPRETATION OF VIBRATIONAL STRUCTURE

Resolved vibrational structure within a band contains useful information in the vibrational intervals themselves, in the identity of the vibrational modes excited and in the relative intensities of the vibrational lines. All of these allow direct deductions to be made about the nature of the ionic state produced, and also about differences between the ionic state and the molecular ground state from which ionization takes place. The changes in molecular geometry caused by ionization of particular electrons can be deduced if the vibrational structure of a band is resolved; they provide the closest characterization of the bonding power of electrons in molecules possible by photoelectron spectroscopy.

5.3.1   FREQUENCIES AND ANHARMONICITIES

The change in frequency of a vibration on ionization is related to the change in bond strength and thus to the bonding power of the electron removed, and little more information can normally be obtained from vibrational spacings alone. One exception is the

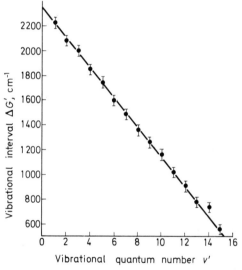

Figure 5.3. Graph of the vibrational interval in H_2^+ ($X\,^2\Sigma_g^+$) against the upper state vibrational quantum number, showing positive anharmonicity. (From Cornford, A. B., Frost, D. C., McDowell, C. A., Ragle, J. L. and Stenhouse, A. I., in Quayle, A. (Editor) *Advances in Mass Spectrometry*, Vol. 5, Institute of Petroleum, London (1971))

halving of the frequency of a bending vibration, which is observed when the molecular geometry changes from bent or non-planar in the molecules to linear or planar in the ions (see Chapter 6, Section 6.5). When a long vibrational progression is found, it is also possible to look for a variation of the spacing of the levels with the vibrational

quantum number, and so to derive the anharmonicity constant, x_{ii}.
The energy of a vibrational level above the vibrationless ground
state ($v = 0$) for excitation of a single oscillator, i, can be written

$$G_{(v \geqslant 1)} = hv_i v_i + hx_{ii} v_i v_i^2 \qquad (5.3)$$

where v_i is the frequency* and v_i the vibrational quantum number.
For the spacing between adjacent lines

$$\Delta G = hv_i + hx_{ii} v_i (2v_i - 1) \qquad (5.4)$$

Hence the interval ΔG should be a linear function of the vibrational
quantum number v_i with a slope of $2hx_{ii} v_i$. If the spacing increases
with increasing v_i, the anharmonicity is said to be negative, although

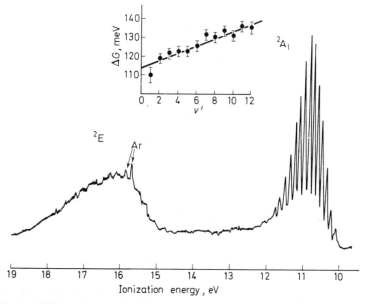

Figure 5.4. Photoelectron spectrum of ammonia with an insert showing negative
anharmonicity in the $^2A_1$ band

the anharmonicity constant, x_{ii}, is positive. The nomenclature arises
because the potential energy curves for most diatomic molecules
lead to negative values of x_{ii}, showing that the levels converge as
they approach the dissociation limit, and such convergence is
considered to be normal, thus positive. Converging progressions

* Strictly this should be written as v_i^o and the anharmonicity constant as x_{ii}^o,
since they are not exactly the same as the constants for infinitesimal amplitude, the
zero-order constants.

are often found in the photoelectron spectra of diatomic molecules; *Figure 5.3* is a graph of ΔG against v' for H_2^+ as an example. Negative anharmonicity is found most commonly in progressions due to bending vibrations; this is because in such vibrations the restoring force does not become weaker as the amplitude increases, but is rather increased by steric interactions when the oscillating groups approach each other. The second bands in the spectra of H_2O, H_2S, H_2Se and H_2Te are examples of this, and the first bands in the spectra of NH_3 and related molecules show the same effect. The increase in the vibrational interval at high vibrational quantum numbers can be seen in the photoelectron spectrum of ammonia in *Figure 5.4*.

5.3.2   IDENTITY OF THE VIBRATIONS EXCITED

If a molecule has several different totally symmetric vibrational modes, a relationship should exist between the localization of bonding or antibonding character in the electron removed and the identity of the vibrational mode that is most strongly excited on ionization. The best examples of a clear relationship of this type

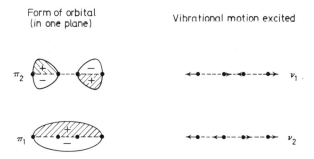

Figure 5.5. The forms of the π orbitals in cyanogen and the vibrational motions excited by ionization from them. The same relationship between orbital structure and motions of the heavy atoms holds for diacetylene, and a similar relationship for the dihaloacetylenes

are found in the photoelectron spectra of the linear molecules of cyanogen[7], diacetylene[7] and the dihaloacetylenes[8]. In all of these molecules, ionization of the outermost π electron is accompanied by excitation of v_1 and ionization of the inner π electron by excitation of v_2. The relationship between the vibrational modes and the orbital structures is shown in *Figure 5.5*.

The outermost π orbitals are bonding across the triple bonds and antibonding across the central single bond; accordingly, the mode excited on ionization, v_1, is one in which the triple bonds contract while the single bond expands. The inner π orbitals are bonding over the whole molecule, but most strongly over the central single bond, and ionization from them excites the symmetrical stretching vibration v_2 in which the central single bond expands most. More often, the type of vibration excited, whether it is, for instance, a bending or a C–C or a C–H stretching, gives an indication of the nature of the electron removed. Ionization from the non-bonding but angle-determining lone-pair electrons in water or ammonia causes strong excitation of the bending vibrations, while removal of electrons that are more strongly involved in the bonding causes excitation of both bending and stretching vibrations. Some further examples from the photoelectron spectrum of benzene have been mentioned in Chapter 4, where the observation of progressions showing excitation of C–H as well as C–C stretching vibrations was used to help assign some of the bands.

5.3.3 CHANGES IN MOLECULAR GEOMETRY ON IONIZATION

Once the vibrational modes excited in a particular ionization have been identified, one can proceed to an interpretation in terms of the changes produced in the molecules; these changes also reflect the bonding character of the electron removed. In diatomic molecules, all is simple. The measured frequencies in different states of the molecule or ion give the bond force constants, while the intensity distributions in the vibrational progressions lead to a determination of the changes in equilibrium bond length. In larger molecules, this simplicity is largely lost because in each normal mode of vibration the frequency is determined by several force constants between different atoms, and the extent of excitation of a particular mode on ionization depends on changes in equilibrium length in several bonds and also on changes in bond angles.

The simplest method of deriving changes in molecular geometry from vibrational line intensities is a direct application of the semi-classical Franck–Condon approximation to bands in which only a single vibrational mode is excited, where the motion in that mode is well described by variations in a single molecular co-ordinate, a bond length or angle. This is automatically true of the diatomic molecules and it is also true of the totally symmetric stretching vibrations in molecules of the forms linear BAB, tetrahedral AB_4, octahedral AB_6, square planar AB_4 and planar AB_3. Bending vibrations of non-linear BAB molecules and the umbrella bending vibration of pyramidal AB_3 molecules are also totally symmetric and can be reasonably described by a single molecular co-ordinate, so the changes in angle following ionization can be represented on a two-dimensional potential energy diagram like that for a diatomic molecule.

Consider the two-dimensional potential energy diagram of *Figure 1.5* with vibrational wave-functions illustrating the Franck–Condon principle. The wave-functions for high vibrational quantum numbers have their greatest amplitude near the potential energy curve itself, which gives the position of the turning points of the classical vibrational motion. The largest Franck–Condon factor for a transition from the ground state, and therefore the maximum intensity in a vibrational progression, occurs when the turning point in the ionic state comes at the same bond distance or angle as the equilibrium position in the ground state of the molecule. The energy of the classical oscillator that represents vibrations in the ion is proportional to the square of the vibrational amplitude at the turning point and can be set equal to the experimental vibrational energy in the ion at the vertical ionization potential. For a stretching vibration in a diatomic molecule.

$$2\pi^2 \mu v'^2 \delta^2 = (v'_{max.} + \tfrac{1}{2})hv' = I_{vert.} - I_{adb.} \qquad (5.5)$$

where v' is the vibrational frequency in the ion, μ is the reduced mass appropriate for the vibration, $v'_{max.}$ is the vibrational quantum number at the maximum intensity in the band and δ is the difference in equilibrium bond length between ion and molecule. The same equation is valid for bending vibrations if δ is replaced by $l\delta\theta$, where l is the bond length and $\delta\theta$ the change in equilibrium bond angle. When the constants and parameters are expressed in convenient units, the equation reduces to

$$\delta^2 = l^2\delta\theta^2 = 5.439 \times 10^5 \frac{(I_{vert.} - I_{adb.})}{\mu v'^2} \qquad (5.6)$$

with δ and l in ångstroms, $\delta\theta$ in radians, $I_{vert.}$ and $I_{adb.}$ in electron volts, μ in atomic units and v' in cm^{-1}.

Because the similarity between the classical and wave-mechanical oscillators used in deriving this equation exists only for high vibrational quantum numbers, the equation is valid only when large changes of shape and therefore long vibrational progressions are involved. Care must be taken when using it to define the reduced mass μ correctly for the particular mode excited, and on this point Herzberg's book[2] must usually be consulted. The energy of the classical vibrator on the left-hand side of equation 5.5 must also be calculated explicitly in terms of the desired internal co-ordinate whenever a triatomic or larger molecule is involved.

A more general method of determining changes in molecular shape from experimental vibrational intensities in electronic transitions has been used by Heilbronner, Muszkat and Schaublin[10]. It is based on calculated Franck–Condon factors for transitions between two harmonic oscillators, as these can be expressed in analytical form. Even when the (0–0) band is the strongest, or if several modes are excited, the measured vibrational intensities and the changes in vibrational frequency between molecule and ion lead directly to determinations of the changes in normal co-ordinates, the co-ordinates that describe the motion in the normal modes of vibration. The changes in normal co-ordinates must then be translated into changes in internal co-ordinates such as bond lengths and bond angles, and in general this is carried out by methods based on matrix algebra[10, 11]. When only one vibration and one molecular parameter are involved, however, the changes caused by ionization can be calculated directly; some bond length changes estimated by this method are given in *Table 5.1* for comparison with

Table 5.1 STRUCTURAL CHANGES FOLLOWING IONIZATION

Molecule	Orbital ionized	Ionic state	Change in bond length on ionization, Å		
			(1)	(2)	(3)
H_2	σ_g (b)	X $^2\Sigma_g^+$	0.33	0.35	0.32
O_2	π_g (ab)	X $^2\Pi_g$	−0.08	−0.09	−0.08
	π_u (b)	a $^4\Pi_u$	0.17	0.20	0.19
CO	σ (nb)	X $^2\Sigma^+$	0.04	0.05	—
	π (b)	A $^2\Pi$	0.12	0.12	0.11
CO_2	π_g (nb)	X $^2\Pi_g$	0.015	0.015	—
	π_u (b)	A $^2\Pi_u$	0.066	0.074	0.062
	σ_u (b)	B $^2\Sigma_u$	0.018	0.015	—
CS_2	π_g (nb)	X $^2\Pi_g$	0.01	—	—
	π_u (b)	A $^2\Pi_u$	—	0.08	0.07

The changes in bond length (r_{AB} for the triatomic molecules) given in the last three columns are derived (1) from rotational analysis of optical spectra, (2) from photoelectron spectra by the method of Heilbronner, Muszkat and Schaublin[10], and (3) by using equation 5.6. For the BAB triatomic molecules, the energy of the classical oscillator, expressed in terms of the A–B bond length, is $4\pi^2\mu\nu'^2\delta_{AB}^2$.

the values obtained by using equation 5.6 and with the accurate bond length changes deduced from the analysis of rotational structure in emission spectra. No method based on Franck–Condon factors can determine the sign of the changes in molecular parameters directly; the direction of change must be deduced from additional evidence, usually from the direction of the change in vibrational frequency following ionization.

The agreement shown in *Table 5.1* between bond length changes determined by the different methods is generally good, and suggests that vibrational intensity measurements could provide a reliable way of finding molecular geometries in ionic states and of determining the bonding powers of different electrons in molecules quantitatively. There are some discrepancies, however, which are due partly to the simplifying assumptions made in the derivations, and perhaps partly to physical effects, such as autoionization, which can make relative vibrational intensities deviate from the calculated Franck–Condon factors. Of the approximations involved, the neglect of anharmonicity may be serious, and accurate changes in molecular parameters must be derived by a different approach. A model is made of the ionic potential energy surface, including all available information on anharmonicities, with the bond lengths and angles as adjustable parameters. Franck–Condon factors for transitions to the state represented by this surface can then be calculated numerically for comparison with the observed spectrum and the parameters adjusted until agreement is reached. This method has been used to calculate Franck–Condon factors for the photoionization bands of water, ammonia and nitrous oxide[12, 13], and so to derive the changes in molecular shape.

5.4 UNRESOLVED BANDS

Most photoelectron spectra of molecules larger than triatomic contain some, if not a majority, of bands that have no apparent vibrational structure. Although theoretical reasons for this lack of structure are discussed below, very common practical reasons are undoubtedly poor instrumental resolution or a poor signal to noise ratio. The positive identification of a continuous band can be made, paradoxically, only on the basis of much more detailed measurements than are needed to find the vibrational structure in a well resolved band. A resolution of 10 meV or better is necessary to be sure of observing vibrational lines separately, and a good ratio of signal to statistical noise is also essential.

5.4.1 THE EXISTENCE OF CONTINUOUS BANDS

Reasons often suggested for the occurrence of an unresolved band are as follows.

(1) The band consists of many closely spaced vibrational lines that cannot possibly be resolved. This requires a vibrational

spacing less than the width of the lines due to their rotational structure, which is of the order of kT (24 meV) in many instances. or less in others where the rotational selection rule is $\Delta J = 0$, ± 2. The broadening due to thermal motion of the target molecules, so important for H_2, is much less so for the larger molecules with which we are concerned here. Vibrations with such low frequencies as 24 meV (200 cm^{-1}) or lower are common in molecules that contain heavy atoms, but much rarer in hydrocarbons, where only torsional vibrations have such frequencies. If only totally symmetric non-degenerate vibrations are considered, frequencies as low as 200 cm^{-1} are unusual unless atoms from the third row of the Periodic Table are involved. If unresolved bands found in the spectra of compounds of first-row elements are to be attributed to overlapping, simultaneous excitation of several

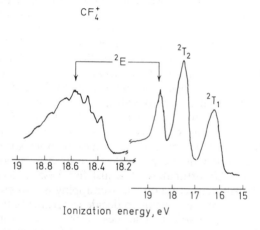

Figure 5.6. The photoelectron spectrum of carbon tetrafluoride, with the ^2E band on an expanded scale so as to show the vibrational structure. (From Pullen, W. E., et al., Inorg. Chem., 9, 2474 (1970), by courtesy of the American Chemical Society)

modes of vibration must probably be invoked. If the totally symmetric vibrations are of such frequencies that they ought to be resolved, a possible reason for unresolved bands in the spectra of symmetrical molecules is the occurrence of Jahn–Teller complications, which indicate a degenerate electronic state. The first two bands in the photoelectron spectrum of carbon tetrafluoride have no resolved vibrational structure (*Figure 5.6*), although the third band is structured, and the only totally symmetric vibrational mode has a frequency of

$904 \, cm^{-1}$ ($112 \, meV$) in the neutral molecule. The reduction in frequency on ionization could hardly be so great as to prevent resolution of such large quanta, so some explanation is needed. A Jahn–Teller effect in the ^2T states corresponding to the first two bands is indeed expected, but another explanation of the continuous nature of the bands is possible, even likely, namely dissociation.

(2) The lifetime of the molecular ions in a particular state is so short that a broadening of the energy levels results from the uncertainty principle, $\Delta E \Delta t \approx \hbar$; alternatively, the ionic state may be repulsive and so truly continuous in the Franck–Condon region. The extent of the broadening necessary to produce a continuous band depends on the spacing of the vibrational structure that would otherwise be expected, and this spacing will normally be more than $20 \, meV$. The uncertainty principle, expressed in units of electronvolts and seconds, is approximately $\Delta E \Delta t = 10^{-15}$, so a broadening of $10 \, meV$ or more corresponds to a lifetime of $10^{-13} \, s$ or less. The lifetimes of the ions produced in ultraviolet photoelectron spectroscopy before radiating, the fluorescence lifetimes, are much longer than this time, about $10^{-8} \, s$, but lifetimes as short as $10^{-13} \, s$ towards decomposition by direct dissociation or pre-dissociation are possible, and perhaps also towards internal conversions to different ionic states (see Chapter 7). Large molecular ions have many possible decomposition pathways and on this account bands without structure are to be expected in the photoelectron spectra of polyatomic molecules. Unfortunately, it is just for these molecules that loss of visible structure due to overlapping is also most likely, and it is therefore difficult to decide experimentally between the two explanations. The best established examples of broadening due to dissociation are all found in the photoelectron spectra of small molecules—for instance, in the $c \, ^4\Sigma_u^-$ state of O_2^+, the A $^2\Sigma^+$ states of HF and HBr and the B $^2\Sigma^+$ state of HCN^+. Some photoelectron bands showing dissociative broadening are illustrated in *Figure 7.3* (p. 177).

There is a statistical theory for the unimolecular decomposition rates of large molecular ions, the quasi-equilibrium theory, one of whose fundamental tenets is that all electronically excited states of the ions relax by internal conversions to the ionic ground state before dissociating. This implies a rule that if loss of structure in a particular band in a photoelectron spectrum is due to rapid dissociation, then all bands at higher ionization potential must be equally continuous. The ions that correspond to higher bands have higher internal excitation energies, and according to the theory they will decompose faster. On this basis, the explanation that the loss of structure in the first photoelectron band of CF_4 is due to fast dissociation is precluded by the presence of resolved structure in the third band. However, there is in fact independent evidence from photoelectron–photoion coincidence spectroscopy (Chapter 7, Section 7.7) that CF_4^+ ions in their ground state do dissociate directly, and probably very rapidly. There are many other molecules whose photoelectron spectra contain continuous bands followed by well resolved bands at higher energy, such as benzene, furan, acetone, methane and neopentane. In seeking the origin of continuous bands in the spectra of such molecules, the above rule should at least be borne in mind. At present, it seems as likely that evidence from photoelectron spectra will cast doubt on the quasi-equilibrium theory as that the validity of the rule will be confirmed.

5.4.2 THE SHAPES OF UNRESOLVED BANDS

Even if no vibrational structure can be seen for physical or experimental reasons, the shapes of bands still give a little information. One can distinguish three types of unresolved band shape, which have been designated[14] types 1, 2 and 3 and which are shown in *Figure 5.7*. In type 1 bands, the low ionization potential edge is

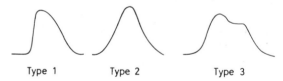

Type 1   Type 2   Type 3

Figure 5.7. Idealized contours of unresolved bands

sharp, indicating that the (0–0) transition is still the strongest, or at least its intensity is substantial, say 50% or more of the most intense transition in the band. Type 2 bands may be roughly symmetrical, but their characteristic feature is the lack of any sharp onset. The FCF for the (0–0) transition is very low, and probably this adiabatic transition is not seen at all. Bands of type 3 have envelopes that are distorted by the Jahn–Teller effect and indicate an orbitally degenerate electronic state.

In bands of type 1, the adiabatic ionization transition can still be discerned as well as the vertical ionization transition, and the difference between vertical and adiabatic ionization potentials is a crude measure of the degree of distortion of the molecule caused by ionization. The fact that the adiabatic transition is seen at all suggests that the distortion is relatively small, and one might therefore expect that within one photoelectron spectrum bands of type 1 would be narrower than bands of type 2. This is often true, and exceptions must be examined carefully with the possibility of overlapping bands in mind. In bands of type 2, however, the widths are determined by different factors from those that affect the difference between vertical and adiabatic ionization potentials. The width of a type 2 band is the product of the extension of the vibrational wave-function in the ground state molecule with the mean slope of the potential energy surface in the ionic state. Neither of these quantities is known beforehand, and the slope of the ionic potential energy surface is a poor indication of the change in molecular geometry. In a polyatomic molecule, both the width of the ground state wave-function and the slope of the potential energy surface in the ion depend on the identity of the normal mode(s) involved. By assuming that the potential energy surfaces are parabolic (harmonic oscillators) and that a single mode is excited, one can derive an equation for the width, Δ:

$$\Delta \propto k^{\frac{1}{2}}\delta/\mu^{\frac{1}{2}} \tag{5.8}$$

where k is the force constant, assumed to be the same in molecule and ion, μ is the reduced mass and δ is the change in the normal co-ordinate. The derivation of this equation involves such severe approximations that its use in estimating δ from the widths of bands of type 2 cannot be recommended. Its important feature is the inverse dependence of band width on $\mu^{\frac{1}{2}}$, which should be observable as an effect of isotopic substitution, particularly deuteration.

5.5   ORBITAL BONDING CHARACTER

The qualitative relationship between photoelectron band form and electron bonding character seems to be very simple at first sight. Long vibrational progressions, broad bands and bands of type 2 and 3 indicate strong bonding or antibonding character, while sharp bands and bands of type 1 indicate little bonding character. Where vibrational structure is resolved, these ideas can be made quantitative by estimation of the direction and magnitude of changes in bond lengths and bond angles following a particular ionization. However, bonding character indicated in this direct way by the photoelectron spectra does not correspond completely to the chemist's normal idea of bonding and non-bonding electrons.

First of all, there are lone-pair electrons, which are non-bonding in the sense that their presence does not affect any bond lengths, but which are angle determining. This is true of the lone-pair electrons in ammonia, amines and other Group V compounds and of the one lone pair of electrons in water, alcohols, ethers and the related Group VI compounds. Ionization from orbitals that have such angle-determining properties gives broad bands in the spectra, often of type 2, even if it causes no change in bond length. Furthermore, non-bonded interactions between lone-pair electrons can drastically alter the effective bonding power of the electrons (see Chapter 8, Section 8.3).

A second problem arises when one considers only molecular orbital theory and covalent bonds, and therefore judges bonding character on the basis of orbital overlap. On this basis, an electrovalent diatomic molecule has no bonding orbitals, as all of the

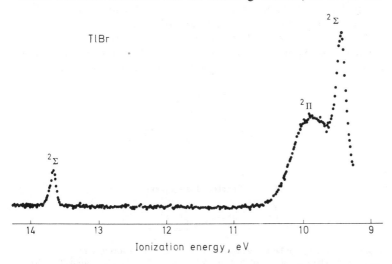

Figure 5.8. He I photoelectron spectrum of thallium bromide. (From Berkowitz[16], by courtesy of the American Institute of Physics)

electrons are located on one or other atom. However, ionization of one of these localized electrons, which converts the molecule from A^+B^- to A^+B, will certainly have a strong effect on the bond length and give a broad ionization band. On the other hand, ionization from a molecular orbital that is equally localized on the two atoms, and is bonding according to a molecular orbital model, will have little effect on the bond length in a predominantly electrovalent molecule. The first example of such a reversal of the normal ideas of bonding character was found in the photoelectron spectrum of xenon difluoride[15] and even clearer examples have been found in the spectra of the thallium and indium monohalides[16, 17]. The spectrum of thallium bromide is shown in *Figure 5.8*, where it can be seen that the low ionization potential region contains only one relatively sharp band and one broad band with which it overlaps. On the normal molecular orbital model, we expect ionization from a lone-pair orbital of the halogen atom to produce a $^2\Pi$ state and give two sharp peaks in this region for the $^2\Pi_{\frac{3}{2}}$ and $^2\Pi_{\frac{1}{2}}$ components, which should be well separated. In fact the *broad* band in the spectrum represents the $^2\Pi$ states, the reason being that the molecules are initially mostly Tl^+Br^- and removal of an electron localized on the halogen has a strong effect on the bonding. The single narrow peak is due to ionization of a σ electron, which is localized about equally on each atom.

Finally, another apparent contradiction of the usual classification of orbitals as bonding or non-bonding is found in the photoelectron spectra of molecules with multiple bonds. According to the photoelectron spectrum of nitrogen (*Figure 1.4*), the $2p\sigma$ orbital is non-bonding, because $2p\sigma^{-1}$ ionization gives a sharp single peak with

Normal p − σ bonding

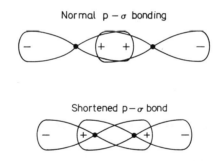

Shortened p − σ bond

Figure 5.9. The loss of bonding power in $p\sigma$ orbitals much shortened by simultaneous π bonding. The elongation of p orbitals has been greatly exaggerated in order to emphasize the less favourable overlap that occurs at short bond lengths

little structure showing vibrational excitation of the ion. This contradicts the simple molecular orbital model in which $2p\sigma$ is the main σ bonding orbital, which contributes with the two π_u orbitals to the triple bond strength of N_2. The reason for the discrepancy is illustrated in *Figure 5.9*. The presence of the π bonds makes the N–N bond so short that positive and negative overlap contributions in $2p\sigma$ cancel out, making the orbital non-bonding at the molecular internuclear distance. The occupancy of $2p\sigma$ does not contribute, therefore, to the force constant (the curvature of the potential energy curve) near the equilibrium internuclear distance, nor does it affect the bond length. It does contribute to the bond dissociation energy, however, because in dissociation the molecule must pass through the region in which the $2p\sigma$ orbital regains its full bonding character. Apparent loss of bonding power such as this is a feature of the photoelectron spectra of all small molecules with multiple bonds; another good example is the σ_u^{-1} ionization in CO_2, which gives a very narrow band and causes very little change in bond length (*Table 5.1*).

REFERENCES

1. SCHWARTZ, M. E., SWITALSKI, J. D. and STRONSKI, R. E., in Shirley, D. A. (Editor) *Electron Spectroscopy*, North Holland, Amsterdam, 605 (1972)
2. COTTON, F. A., *Chemical Applications of Group Theory*, Interscience, New York (1963); HERZBERG, G., *Infrared and Raman Spectra*, Van Nostrand, Princeton, N.J. (1945)
3. TURNER, D. W., *Phil. Trans. R. Soc., Lond.*, **A268**, 7 (1970)
4. BRUNDLE, C. R. and TURNER, D. W., *Int. J. Mass Spectrom. Ion Phys.*, **2**, 195 (1969)
5. ELAND, J. H. D. and DANBY, C. J., *Int. J. Mass Spectrom. Ion Phys.*, **1**, 111 (1968)
6. BRANTON, G. R., FROST, D. C., MAKITA, T., MCDOWELL, C. A. and STENHOUSE, I. A., *Phil. Trans. R. Soc., Lond.*, **A268**, 77 (1970)
7. BAKER, C. and TURNER, D. W., *Proc. R. Soc., Lond.*, **A308**, 19 (1968)
8. HEILBRONNER, E., HORNUNG, V. and KLOSTER-JENSEN, E., *Helv. Chim. Acta*, **53**, 331 (1970)
9. POTTS, A. W. and PRICE, W. C., *Proc. R. Soc., Lond.*, **A326**, 165 (1971)
10. HEILBRONNER, E., MUSZKAT, K. A. and SCHAUBLIN, J., *Helv. Chim. Acta*, **54**, 58 (1971)
11. SMITH, W. L. and WARSOP, P. A., *Trans. Faraday Soc.*, **64**, 1165 (1968)
12. BOTTER, R. and ROSENSTOCK, H. M., *J. Res. natn. Bur. Stand.*, **73A**, 313 (1969)
13. ROSENSTOCK, H. M., *Int. J. Mass Spectrom. Ion Phys.*, **7**, 33 (1971)
14. GLEITER, R., HEILBRONNER, E. and HORNUNG, V., *Helv. Chim. Acta*, **55**, 255 (1972)
15. BREHM, B., MENZINGER, M. and ZORN, C., *Can. J. Chem.*, **48**, 3193 (1970)
16. BERKOWITZ, J., *J. chem. Phys.*, **56**, 2766 (1972)
17. BERKOWITZ, J. and DEHMER, J. L., *J. chem. Phys.*, **57**, 3194 (1972)

6 Photoelectron Band Structure—II: Degenerate Ionic States

6.1 INTRODUCTION

Whenever an electron is ejected from a fully occupied degenerate orbital in a molecule, the result is an orbitally degenerate doublet state of the corresponding ion. The degeneracy of such a state can be lifted either by coupling between the spin and orbital angular momenta of the unpaired electron, spin–orbit coupling, or by a change in molecular shape, the Jahn–Teller effect. When both of these effects are weak, the photoelectron spectrum contains a single intense band corresponding to the ionization, but the vibrational structure of this band is liable to be complex. When the interactions are stronger there can be as many photoelectron bands in the spectrum as there were electron pairs in the original degenerate orbital, but not more. Orbitally degenerate ionic states can also arise from photoionization out of partially occupied degenerate orbitals, from two-electron transitions or from ionization out of closed shells in molecules that also have an open shell. Orbital degeneracy arising from these processes is not common in photoelectron spectroscopy, but whatever the source of degeneracy, the same splitting mechanisms operate. In some unusual photoionization processes, orbitally degenerate ionic states can be produced that are not also spin degenerate; such states would be susceptible to the Jahn–Teller effect only and not to splitting by spin–orbit coupling. For linear species, on the other hand, only spin–orbit coupling can lift the degeneracy and no Jahn–Teller effect is operative.

6.2   SPIN–ORBIT COUPLING

If an unpaired electron is in a degenerate orbital where it has orbital angular momentum, the spin angular momentum and orbital angular momentum can combine in different ways and produce new states that are characterized by the total electronic angular momentum. The new states have different energies because the magnetic moments due to electron spin and orbital motion may oppose or reinforce one another. All states with multiplicity greater than one and a non-zero orbital angular momentum are split by this spin–orbit coupling. The simplest examples from photoelectron spectra are the ionizations of the rare gases, which leave ions with the configuration np^5 in the outermost shell and hence in ^2P states, which split into $^2P_{\frac{3}{2}}$ and $^2P_{\frac{1}{2}}$. Because in this instance the incomplete shell is more than half full, the $^2P_{\frac{3}{2}}$ state, with the higher total angular momentum, is of lower energy than $^2P_{\frac{1}{2}}$, as indicated by Hund's rules[1]. For atomic ions such as these, the degeneracies are equal to $2J+1$, i.e., 4 and 2, respectively, which leads to a $2:1$ ratio of peak intensities in the spectra.

Table 6.1 SPLITTINGS OF DEGENERATE STATES

Point group	Electronic state	Spin–orbit components	Examples	Spin–orbit splitting	Jahn–Teller active modes
C_{3v}	^2E	$E_{\frac{3}{2}} + E_{\frac{1}{2}}$	CH_3I^+	ζ_I	e
C_{4v}	^2E	$E_{\frac{3}{2}} + E_{\frac{1}{2}}$	BrF_5^+		$b_1 + b_2$
$D_{\infty h}$	$^2\Pi_g$	$E_{\frac{3}{2}g} + E_{\frac{1}{2}g}$	HgI_2^+	ζ_I	None
	$^2\Pi_u$	$E_{\frac{3}{2}u} + E_{\frac{1}{2}u}$	XeF_2^+	$c_{Xe}^2 \zeta_{Xe}$	None
$C_{\infty v}$	$^2\Pi$	$E_{\frac{3}{2}} + E_{\frac{1}{2}}$	N_2O^+		None
D_{3h}	$^2E'$	$E_{\frac{3}{2}} + E_{\frac{1}{2}}$	BBr_3^+	ζ_{Br}	e'
	$^2E''$	$E_{\frac{3}{2}} + E_{\frac{1}{2}}$	BBr_3^+	Zero	e'
D_{6h}	$^2E_1$	$E_{\frac{3}{2}} + E_{\frac{1}{2}}$	$C_6H_6^+$	Zero in	e_{2g}
	$^2E_2$	$E_{\frac{3}{2}} + E_{\frac{1}{2}}$	$C_6H_6^+$	$^2E_{1g}, ^2E_{2u}$	e_{2g}
T_d	^2E	$G_{\frac{3}{2}}$	CBr_4^+	(None)	e
	$^2T_1$	$G_{\frac{3}{2}} + E_{\frac{1}{2}}$	CBr_4^+	$\frac{3}{4}\zeta_{Br}$	e, t_2
	$^2T_2$	$G_{\frac{3}{2}} + E_{\frac{1}{2}}$	CBr_4^+	$\frac{3}{2}\zeta_{Br}$	e, t_2
O_h	^2E	$G_{\frac{3}{2}}$	SF_6^+	(None)	e_g
	$^2T_1$	$G_{\frac{3}{2}} + E_{\frac{1}{2}}$	SF_6^+		e_g, t_{2g}
	$^2T_2$	$G_{\frac{3}{2}} + E_{\frac{1}{2}}$	SF_6^+		e_g, t_{2g}

The third column gives the species of the multiplet components produced by spin–orbit splitting in the degenerate electronic states (second column) of molecular ions belonging to important point groups (first column). The fifth column gives some estimates of the splitting in terms of atomic splitting parameters, ζ_l, appropriate for the examples cited in the fourth column. The last column contains the species of the vibrational modes that are Jahn–Teller active for the same electronic states. The g and u symmetries of the multiplet components for D_{6h} and O_h are omitted, as they are the same as those of the electronic states; only g vibrations are Jahn–Teller active in g or u electronic states alike.

When spin–orbit coupling in molecules is considered, the designation of states by the normal symmetry species is not sufficient. In order to determine the number and species of the states produced by spin–orbit coupling, extended point groups, which include the electron spin explicitly[2], must be used. The way in which the normal symmetry species of doublet orbitally degenerate states go over to species of the extended groups is shown in *Table 6.1* for the most important cases encountered in photoelectron spectroscopy.

6.2.1 LINEAR MOLECULES

In linear molecular ions, all $^2\Pi$, $^2\Delta$, $^2\Phi$ states and, of course, triplet or quartet Π, Δ and Φ states are split; the doublet states give two components characterized by a quantum number Ω equal to $\Lambda \pm S$. The two states are of equal degeneracy because there is only one axis in which angular momentum is quantized, and the two peaks in the spectrum should be of equal intensity. By far the commonest states in photoelectron spectra are doublet states produced by removing one electron from a closed shell, and here the term with higher Ω is of lower energy. If spin–orbit interaction is relatively small and is well described by Russell–Saunders coupling[1], the energies of the spin–orbit components are given by

$$E = E_0 + \zeta\Lambda\Sigma \qquad (6.1)$$

where E_0 is the energy in the absence of spin–orbit coupling and Σ is the component of S, the resultant spin angular momentum, along the axis. As Σ takes the values $S, (S-1) \ldots -S$, in this approximation every term is split into $2S+1$ equally spaced components, the spacing being equal to ζ in a Π state ($\Lambda = 1$). The spin–orbit interaction parameter, ζ, is characteristic of the molecular orbital in which the unpaired electron moves, and of the atomic orbitals from which it is made up. For a single electron in the coulomb field produced by a nucleus of charge Z, the splitting increases as Z/r^3, or, as r is proportional to Z^{-1}, ζ must increase with Z^4. For atoms with many electrons, the splitting in the valence shells is approximately proportional to Z^2. For elements in the first row of the Periodic Table, splittings of 20 meV or less are normal, and for those in the second row the splittings are about 50 meV. Because of the intervening filling of the d shells of the transition elements, non-metals of the third row have splittings of 200–300 meV and in the fourth row the splittings are 300–600 meV. When elements of the third, fourth and fifth rows are involved, spin–orbit splittings become particularly noticeable in photoelectron spectra.

In a $^2\Pi$ or $^2\Delta$ state of a linear ion, the magnitude of the splitting depends on the proportions of the different atomic orbitals that make up the molecular orbital in which the unpaired electron moves. If the molecular orbital ψ is expressed as a linear combination of atomic orbitals, ϕ_μ, for atoms μ, with coefficients c_μ:

$$\psi = \sum_\mu c_\mu \phi_\mu \tag{6.2}$$

then the effective splitting ζ is given approximately by

$$\zeta = \sum_\mu c_\mu^2 \zeta_\mu \tag{6.3}$$

where ζ_μ are the characteristic splittings of the atomic orbitals, a selection of which are given in *Table 6.2*. Equation 6.3, although only approximate, is extremely useful as it enables one to determine the coefficients c_μ from the spin–orbit splittings observed in the spectra.

Table 6.2 ATOMIC SPIN–ORBIT COUPLING PARAMETERS FOR ATOMS AND IONS

Orbital	Splitting parameter ζ, eV						
2p	B 0.001	C 0.004	N$^+$ 0.011	O −0.019	F −0.033	Ne$^+$ −0.064	Na$^+$ −0.121
3p	Al 0.009	Si 0.019	P$^+$ 0.039	S −0.047	Cl −0.073	Ar$^+$ −0.118	K −0.173
4p	Ga 0.068	Ge 0.12	As$^+$ 0.22	Se −0.21	Br −0.305	Kr$^+$ −0.444	Rb −0.563
5p	In 0.182	Sn 0.32	Sb$^+$ (0.38)	Te (0.4)	I −0.628	Xe$^+$ −0.871	Cs −0.8̇0
6p	Tl 0.644	Pb 1.15	Bi$^+$ (1.9)				
3d	Sc 0.008	Ti 0.013	V 0.020	Cr 0.028	Mn 0.030	Fe 0.048	Co 0.064
3d	Ni 0.074	Cu 0.102					

The values for the transition metals are from J. S. Griffith, *The Theory of Transition Metal Ions*, Cambridge University Press (1961), which may also be consulted for the method of deriving these values from atomic spectra. The other values are based on the tables of C. E. Moore, *Atomic Energy Levels*, Vol. I (1949), Vol. II (1952), Vol. III (1957), National Bureau of Standards, Washington, D.C. The ζ values for the C, N$^+$ and O series are based on an assumption of LS coupling, and so must be considered to be approximate. The most doubtful values are given in parentheses.

The simplest examples of spin–orbit splittings are found in the spectra of the halogen hydrides. In these molecules there is only one π orbital, which is wholly localized on the halogen atom, so that ionization leads to $^2\Pi_{\frac{3}{2}}$ and $^2\Pi_{\frac{1}{2}}$ states (strictly $E_{\frac{3}{2}}$ and $E_{\frac{1}{2}}$ in the extended point group), and the splittings between them should be equal to ζ_X. The observed splittings for HCl, HBr and HI given in *Table 6.3* are in excellent agreement with this idea. The spectra of these molecules are particularly simple; they show ionization from the p orbitals of the halogen, two of which form the π orbital while the other forms the H–X σ bond. The bonding character of the pσ orbital is apparent from the vibrational structure of the σ^{-1} ionization bands, which is visible in the photoelectron spectrum of HCl (*Figure 6.1*).

Table 6.3   SPIN–ORBIT SPLITTING IN LINEAR IONS

Dominant atom	One heavy atom		Two heavy atoms		
	Ion	Splitting (eV) in $X\ ^2\Pi$ or ^2E	Ion	Splittings (eV) in	
				$X\ ^2\Pi_{(g)}$	$A\ ^2\Pi_{(u)}$
O			O_2^+	0.024 2	
			CO_2^+	0.019 8	0.011 8
F	HF^+	0.030	F_2^+	0.03	
S			CS_2^+	0.054 6	
Cl	HCl^+	0.080	Cl_2^+	0.08	
			ClF^+	0.078	
			$HgCl_2^+$	0.120	
Br	HBr^+	0.333	Br_2^+	0.35	0.34, 0.22*
	CH_3Br^+	0.315	BrF^+	0.322	
	$(CH_3)_3CBr^+$	0.29	$HgBr_2^+$	0.34	0.34
I	HI^+	0.66	I_2^+	0.65	0.80
	CH_3I^+	0.627	ICl^+	0.58	
	$(CH_3)_3CI^+$	0.56	IBr^+	0.58	0.36
			HgI_2^+	0.66	0.40

6 Value uncertain; see Evans, S. and Orchard, A. F., *Inorg. Chim. Acta*, 5, 81 (1970).

The next examples to be considered are the halogens and binary interhalogen compounds, whose measured spin–orbit splittings are also given in *Table 6.3*. Their photoelectron spectra each contain three bands, reflecting the outer orbital structure of the molecules: $\sigma_g^2\ \pi_u^4\ \pi_g^{*4} \dots {}^1\Sigma_g^+$ (the designations g and u must be omitted for the interhalogens). The spectrum of iodine is shown in *Figure 6.2*. For the outer π orbital ionizations of the halogens the spin–orbit splittings are close to the atomic coupling parameters, as expected according to equation 6.3, but noticeable deviations occur in the inner π orbital ionizations.

The increased splitting in the $^2\Pi_u$ state of I_2^+ is caused by a spin–orbit induced interaction between the π and σ orbitals[3]. In the extended point group, the unoccupied σ_u orbital has the same species, $e_{\frac{1}{2}u}$, as the higher energy component of π_u, so the two orbitals repel each other in the usual manner of interacting orbitals. Similarly, the full σ_g orbital has the same species in the double group as the $\Omega = \frac{1}{2}$ component of π_g, and these orbitals also repel each other. The result is a reduced splitting in $^2\Pi_g$ and an increased splitting in $^2\Pi_u$, as illustrated in *Figure 6.3*. This is a second-order effect, and such effects are strongest when the interacting orbitals have similar energies.

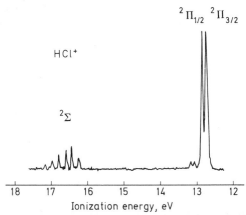

Figure 6.1. Photoelectron spectrum of HCl ionized by He I light

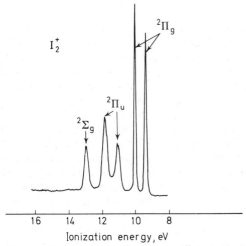

Figure 6.2. Photoelectron spectrum of iodine showing spin–orbit splitting in the $^2\Pi_g$ and $^2\Pi_u$ ionic states. (From Evans, S. and Orchard, A. F., *Inorg. Chim. Acta*, **5**, 81 (1971), by permission)

The splittings observed in the $^2\Pi$ states of the interhalogen ions are reasonable in terms of equation 6.3 and the expected orbital characters, but because of the likely presence of second-order effects in ICl and IBr it is not appropriate to deduce atomic orbital coefficients. With ClF^+, for which the second-order effects should be small, excellent agreement with the observed splittings has been obtained by the use of a more elaborate, but related, theory[4].

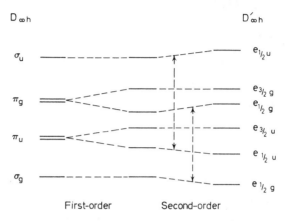

Figure 6.3. First- and second-order spin–orbit interactions in I_2^+ according to Wittel[3]. See also Jungen, M., *Theor. Chim. Acta*, **27**, 33 (1972), for an alternative interpretation

We next consider the 16-electron linear triatomic molecules carbon dioxide, carbon disulphide and the mercury(II) halides. Their valence electronic structure can be written as $\sigma_u^2 \, \sigma_g^2 \, \pi_u^4 \, \pi_g^4 \dots$ $^1\Sigma_g^+$. The orbital π_g is located entirely on the outer atoms and is non-bonding, while π_u is delocalized over the whole molecule with greatest density at the central atom, and is bonding. Because the electrons in π_g are located on the outer atoms, the splittings in the $^2\Pi_g$ states should be equal to the atomic splitting parameters for the outer atoms, as indeed they very nearly are. The splittings in the inner $^2\Pi_u$ states of CO_2^+ and CS_2^+ are considerably smaller, because of the low ζ value for carbon 2p compared with that of oxygen 2p or sulphur 3p. The situation with mercury(II) bromide and iodide is similar, but the characteristic splitting in the Hg 6p orbital is estimated to be large and reduces the splitting in $^2\Pi_u$ compared with $2\Pi_g$ only in HgI_2^+.

Xenon difluoride is also a linear molecule, but has 22 valence electrons. Its molecular orbital structure deduced from the photoelectron spectrum (Chapter 4, Section 4.5.2) is $\sigma_u^2 \, \pi_u^4 \, \pi_g^4 \, \sigma_g^2 \, \pi_u^{*4}$. The outermost occupied orbital, π_u^*, is the antibonding counterpart of

π_u, which in this molecule, as in the previous molecules, is bonding and has its greatest density at the central atom. The π_g orbital is again non-bonding and located on the outer atoms only. The spin–orbit parameter for xenon 5p is much larger than that for fluorine 2p, so that it is in the $^2\Pi_u$ states that large splittings are to be expected. The splitting in the outer $^2\Pi_u$ state of XeF_2^+ is found to be 0.47 eV. Then, from equation 6.3 and the atomic splitting parameter for xenon 5p (0.871 eV), the percentage of xenon 5p character in the π_u^* orbital must be $0.47/0.87 \times 100 = 54\%$. For the bonding and antibonding orbitals π_u and π_u^*, the wave-functions can be written approximately as

$$\psi_{\pi_u^*} = c_1 \phi_{Xe5p} - c_2 \phi_{(F+F)} \tag{6.4}$$

$$\psi_{\pi_u} = c_2 \phi_{Xe5p} + c_1 \phi_{(F+F)} \tag{6.5}$$

The percentage of xenon 5p character in the π_u orbital must be 46% and the spin–orbit splitting is predicted to be 0.4 eV, in exact agreement with the experimental value $(0.40 \pm 0.07 \text{ eV})$.

Other linear molecules whose photoelectron spectra contain $^2\Pi$ bands split by spin–orbit coupling are the mono- and di-halo-acetylenes[5, 6]. For these molecules, the use of equation 6.3 with atomic orbital coefficients obtained from semi-empirical molecular orbital calculations gave excellent agreement with the experimental splittings. The photoelectron spectrum of krypton difluoride has also been measured recently and analysed in a similar way[7].

6.2.2 NON-LINEAR MOLECULES

In non-linear molecules, the electrons have orbital angular momentum other than zero only in degenerate orbitals. Orbital degeneracy based on symmetry can exist only in molecules with at least a three-fold axis, so that the molecular point group contains degenerate irreducible representations. In order to determine the number and species of the new terms produced by spin–orbit coupling in multiplet orbitally degenerate states of non-linear molecules, it is essential to use the extended point groups; the most useful data are given in *Table 6.1*. However, the number and symmetries of the new terms unfortunately give no indication of the magnitude of the spin–orbit splittings, except for the ^2E states of tetrahedral and octahedral molecules where there is only one component and the splitting is zero. When one atom or set of equivalent atomic orbitals has the dominant characteristic splitting, an approximate value of the molecular splitting can sometimes be estimated on a model similar to the one used for linear molecules; some estimates of this type are shown in the last column of *Table 6.1*. One striking entry in this column indicates that in $^2E''$ states of D_{3h} molecules the splitting is identically zero, although two species are generated in the extended group. This effect can be explained in general terms as follows. The splitting in axially symmetric molecules is proportional to the product of the projections of the orbital and spin angular momenta in the symmetry axis. An e'' orbital, in BI_3^+, for instance, is made up by combining out-of-plane (π-type) p orbitals of the iodine atoms, and the axes of these p orbitals are parallel to the three-fold rotation axis. The projection of orbital angular momentum in this axis is identically zero and no spin–orbit splitting can result. The same is true of out-of-plane π orbitals in other planar symmetric molecules, such as benzene. It cannot be concluded, however, that $^2E''$ states are never split by spin–orbit interaction, because second-order effects, as discussed for I_2^+, are very likely to cause splittings, although they are usually smaller than the first-order splittings.

Examples of spin–orbit interactions in non-linear molecular ions are found in the photoelectron spectra of the tetrahedral AB_4 molecules. The photoelectron spectrum of CF_4 (*Figure 5.1*) shows that three electronic states, $^2T_1$, $^2T_2$ and ^2E, arise from ionization out of the halogen lone-pair orbitals, and in these states of CF_4^+ the spin–orbit splitting is negligible. In CBr_4^+, however, substantial splittings are expected, and for comparison the lone-pair ionization region of the spectrum is illustrated in *Figure 6.4*. According to *Table 6.1*, the ^2T states should be split into two components and the ^2E state should remain degenerate, giving a total of five ionic states. The observation of five bands in the CBr_4 spectrum is in agreement with this prediction, and the unsplit ^2E band can be recognized at once. The small additional splittings on two of the five bands are also expected, because the G states produced by spin–orbit coupling are still doubly degenerate and are subject to Jahn–Teller splittings (but not to further spin–orbit splitting). The magnitudes of the experimental energy separations between the spin–orbit split states of CBr_4^+ agree with the estimates in column five of *Table 6.1* and the characteristic splitting parameter, ζ_{Br}, for bromine 4p electrons. In the spectrum of $Pb(CH_3)_4$, which is also shown in *Figure 6.4*, the spin–orbit split band corresponds to ionization from an orbital in which the p orbitals of the Pb atom play a large part. The splitting in this instance into $G_{\frac{3}{2}}$ and $E_{\frac{5}{2}}$ is similar to the splitting of a ^2P state of an atomic ion, and the components should be separated by $\frac{3}{2}\zeta_{Pb}$ if the orbital were purely Pb 6p.

The photoelectron spectrum of BI_3, shown in *Figure 6.5*, provides an example of second-order effects[8]. The first three orbitals from which ionization occurs are of a'_2, e' and e'' symmetry, so that according to *Table 6.1* there should be only four ionization bands showing that the $^2E'$ state is split, but not $^2E''$. In fact, five peaks are seen in the halogen lone-pair ionization region, and their number

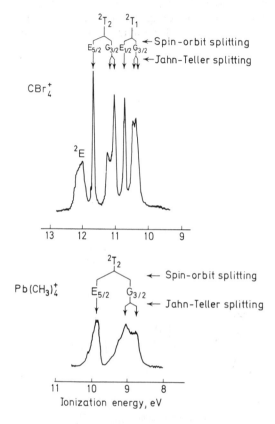

Figure 6.4. Partial photoelectron spectra of carbon tetrabromide and lead tetramethyl showing spin–orbit splitting effects. (CBr$_4$ from Green, J. C., Green, M. L. M., Joachim, P. J., Orchard, A. F. and Turner, D. W., *Phil. Trans. R. Soc., Lond.*, **A268**, 111 (1970), by courtesy of the Council of the Royal Society; Pb(CH$_3$)$_4$ from Evans *et al.*, *J. chem. Soc. Faraday Trans. II*, **68**, 905 (1972), by courtesy of the Chemical Society)

alone shows that the $^2E''$ state is split by a second-order effect. A possible identification of the individual bands, and an indication of the interaction between the E$_{\frac{3}{2}}$ components of $^2E'$ and $^2E''$, are also given in *Figure 6.5*.

The simplest cases of spin–orbit splittings in degenerate states of non-linear molecular ions would appear to be the lone-pair ionizations of the methyl halides. The lone-pair electrons are in e orbitals, so ionization gives ^2E states, which split into two components, $E_{\frac{3}{2}}$ and $E_{\frac{1}{2}}$, of equal statistical weight. There is another e orbital in the molecules made up from carbon 2p orbitals and responsible for the CH bonding, but conjugation of the lone-pair e electrons with it is weak and the magnitude of the splitting should

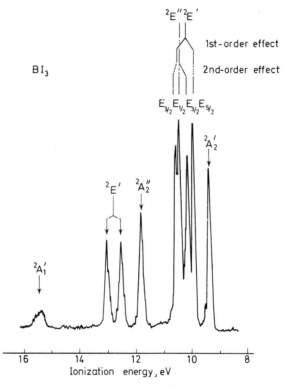

Figure. 6.5. Photoelectron spectrum of boron triiodide, showing a second-order spin–orbit splitting. The analysis indicated is from King et al.[8]

be ζ_X, as in a linear molecule. This simplicity makes it possible to recognize the existence of a new complication. The spectra of methyl iodide and methyl bromide shown in *Figure 6.6* do contain two peaks in the expected region separated by ζ_I or ζ_{Br}, but in the spectrum of methyl bromide their intensities are not equal. This complication arises because the ^2E states can have their degeneracy lifted not only by spin–orbit coupling but also by the Jahn–Teller effect, as is usual. In MeBr$^+$, the Jahn–Teller splitting is smaller than the spin–orbit splitting, but manifests itself by making allowed the excitation of otherwise forbidden vibrations. It happens that one of the vibrationally excited levels of a normally forbidden progression starting from the $E_\frac{3}{2}$ state is of the same symmetry as the vibrationless

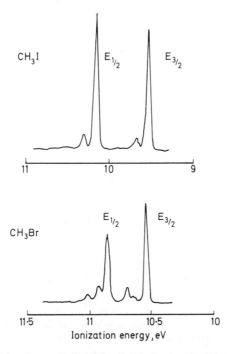

Figure 6.6. Halogen lone-pair ionization bands in the photoelectron spectra of methyl iodide and methyl bromide. (From Ragle, J. L., Stenhouse, I. A., Frost, D. C. and McDowell, C. A., *J. chem. Phys.*, **53**, 178 (1970), by courtesy of the American Institute of Physics)

$E_\frac{1}{2}$ state, and similar in energy. The result is an unnoticeable shifting of the levels and a strong borrowing of intensity from the $E_\frac{1}{2}$ peak into the vibrational structure of the $E_\frac{3}{2}$ peak. If deuterium is substituted for hydrogen, giving CD_3Br, the vibrational frequencies are changed sufficiently to reduce the effect noticeably, and two peaks of almost equal intensity appear in the spectrum for $E_\frac{1}{2}$ and $E_\frac{3}{2}$.

After Jahn–Teller effects have been considered in more detail, the competition between them and spin–orbit splittings is discussed again in Section 6.4.

6.2.3 SPIN–ORBIT COUPLING WITHOUT DEGENERACY

According to the preceding discussion, no spin–orbit splitting should be observed in the ionic states of molecules with less than a three-fold symmetry axis, as such molecules contain no orbitals that are degenerate because of symmetry. This is apparently contradicted by the fact that no discontinuous variation of the photoelectron spectra is found on going from methyl iodide (C_{3v}) to ethyl iodide (C_s), or from isobutyl iodide (C_1) to t-butyl iodide (C_{3v}). In all instances, the spectra contain two sharp peaks at an ionization potential characteristic of iodine and separated by about 0.6 eV, the normal spin–orbit splitting for an iodine ion in a linear environment. The same lack of discontinuity occurs in the series of alkyl bromides. One possible explanation is that the conjugative interactions of the halogen $p_{x, y}$ orbitals with the orbitals of the alkyl moieties are so weak that the halogens experience a cylindrically symmetrical field. This explanation is contradicted, however, by the observation of vibrational structure on the lone-pair peaks, which indicates the existence of some bonding interaction of the halogen $p\pi$ orbitals with the remainder of the molecule. An explanation of this difficulty has been given by Brogli and Heilbronner[9], who showed that the constancy of the splitting is a result of two opposing effects. Conjugation reduces the spin–orbit splitting by progressively reducing the cylindrical symmetry of the field, and also by adding contributions from lighter atoms to the molecular orbitals, but it simultaneously produces a new splitting. If the spin–orbit interaction is zero or very weak compared with the conjugative interaction, there is still a splitting, which is visible, for example, in the photoelectron spectrum of vinyl chloride. This is a splitting between the halogen p orbitals, p_x and p_y, which are made non-equivalent and therefore non-degenerate by the conjugative interaction. In a molecule such as this with C_s or lower symmetry, one halogen p orbital always conjugates more strongly than the other with the remainder of the molecule, so the two p orbital energies are no longer the same. As the conjugative interaction is increased continuously, the reduction in spin–orbit splitting and the increase in the new splitting caused by conjugation compensate for one another until the conjugation becomes very strong. The model proposed by Brogli and Heilbronner[9] also explains why, in the spectrum of cyclopropyl bromide and other molecules in which conjugation is strong, one of the two peaks corresponding to ionization from the lone-pair orbitals has broad vibrational structure while the other is sharp. The sharp peak corresponds to the atomic p orbital of the halogen, which does not interact with the remainder of the molecule, and whether this is the peak at higher or lower ionization potential will depend on the relative energy of the interacting orbitals, as indicated in *Figure 6.7.*

There is a continuous transition from the rather weak conjugation and dominant spin–orbit coupling of the n-alkyl halides to the strong conjugation with one lone-pair orbital of the halogen atom in unsaturated halogen compounds such as vinyl chloride or the halobenzenes. The number of bands observed for ionization from the lone-pair orbitals of a single bromine or iodine atom is always two, equal to the original orbital degeneracy, irrespective of whether

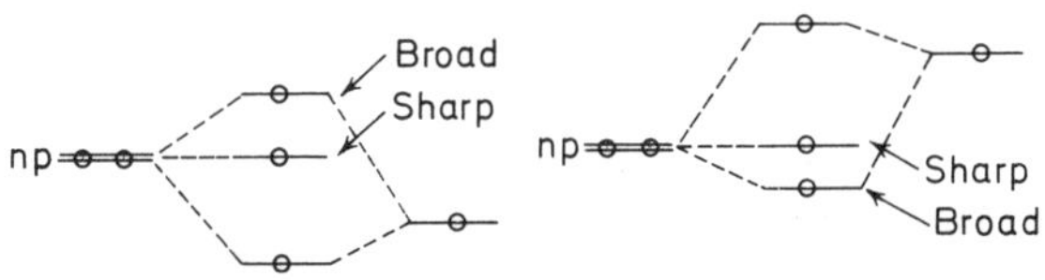

Figure 6.7. Molecular orbital diagrams showing two idealized cases of conjugative interactions between halogen lone-pair p orbitals and other occupied orbitals of a molecule. The designations 'sharp' and 'broad' refer to the characters of the corresponding photoelectron bands

the degeneracy is lifted by spin–orbit coupling, conjugative interaction with molecular orbitals of lower symmetry or, as we shall see below, by Jahn–Teller effects.

Another group of molecules that have too low a symmetry to possess degenerate orbitals is typified by the methylene halides, CH_2X_2. The non-bonding valence electron ionization regions of the photoelectron spectra of methylene chloride and iodide are depicted in *Figure 6.8*; before considering the symmetries, it might have been guessed that the separation of the two bands in the spectrum of CH_2Cl_2 into four in the spectrum of CH_2I_2 was due to spin–orbit splittings. This naïve idea is at least partly correct, because spin–orbit splitting can arise from *accidental* degeneracy as well as symmetry-based degeneracy. In other words, when two orbitals are nearly degenerate, the second-order effects can become as strong as first-order effects. Alternatively, for masters of quantum mechanics, the spin–orbit interaction produces no diagonal matrix elements of the Hamiltonian, but large off-diagonal matrix-elements.

THE JAHN–TELLER EFFECT

The theorem of Jahn and Teller[10] states that a non-linear molecule in a degenerate electronic state is unstable towards distortions which remove the degeneracy. Although this applies in general to both spin and orbital degeneracy, the effects due to spin are always negligibly small, and we are concerned in practice with orbitally degenerate states. Whenever such a state is produced by ionization of a non-linear molecule, usually by removal of one electron from a degenerate orbital, the positive ion may distort to a lower symmetry, thereby

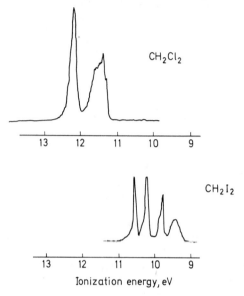

Figure 6.8. Halogen lone-pair ionization bands in the photoelectron spectra of methylene chloride and iodide

becoming more stable. The distortion is brought about by, or is equivalent to, excitation of one or more degenerate* vibrational modes of the undistorted molecule, which are called the Jahn–Teller active vibrations. The species of the active vibrations for molecules of all important point groups have been tabulated by Herzberg[2] and some are reproduced in the final column of *Table 6.1*. As the excitation of these vibrations causes a change in electronic energy, the total energy can no longer be separated into electronic and vibrational parts; that is, there is strong vibronic coupling. The wave-functions also can no longer be separated so all the properties of the system must strictly be considered in terms of the vibronic levels that result from the coupling of vibrational and electronic

* In a square-planar molecule, the Jahn–Teller active vibrations are not degenerate but are the two simple vibrations of species b_{1g} and b_{2g}. These act together in exactly the same way as the two components of a degenerate vibration.

motion. Nevertheless, it is helpful to distinguish between changes in the effective potential energy surfaces, referred to as the *static* Jahn–Teller effect, and complications of the vibrational (vibronic) structure of the state, called the *dynamic* Jahn–Teller effect.

The strength of the Jahn–Teller effect is measured in terms of a dimensionless parameter D, which is so defined that the product of the frequency of the active vibration with D and Planck's constant is equal to the Jahn–Teller stabilization energy, the reduction in energy on going from the symmetrical nuclear configuration to the new equilibrium positions. There is not just one but several new equilibrium positions, and the total original symmetry of the molecule is retained when all the new equilibrium positions are taken together. This concept can be illustrated by the hypothetical case of methyl chloride, a C_{3v} molecule, in an ionic state with a strong Jahn–Teller effect. In this ^2E state of CH_3Cl^+, the equilibrium position of the chlorine atom is no longer on the symmetry axis of the CH_3 group as v_6 is the active vibration. There are three equivalent positions of minimum energy through which the chlorine atom can pass if it is moved around, but not on, the axis of the CH_3 group. In each single position the original symmetry of the molecule is lost, but the system of all three together still has the full C_{3v} symmetry. The Jahn–Teller stabilization energy, hvD, is the difference between the energies of the minima and the symmetrical position, and if D is much less than unity the zero-point energy in the active vibration will be sufficient to take the system from one minimum to another. The photoelectron spectrum will then contain no direct evidence of the new potential energy surfaces but only disturbance of the vibrational structure, and it can be said that there is no static effect but only a dynamic Jahn–Teller effect. Different authorities define the static and dynamic Jahn–Teller effects in slightly different ways[2, 11], but the observational distinction given here is sufficient for the present purpose.

6.3.1   JAHN–TELLER EFFECTS IN PHOTOELECTRON SPECTRA

The dynamic Jahn–Teller effect involves splitting of the vibrational levels by vibronic interactions when active vibrations are excited, but this splitting is not easily observed in photoelectron spectroscopy. The simple presence of progressions that show excitation of the Jahn–Teller active modes is much more noticeable, and is the most sensitive indication of the existence of Jahn–Teller effects. In the absence of vibronic coupling, the excitation of these modes is effectively forbidden (see Chapter 5, Section 5.2.1) because they are degenerate or antisymmetric. Their excitation under the conditions of the Jahn–Teller effect can be understood in two ways. Firstly, selection rules based on vibronic wave-functions instead of the vibrational overlap integrals must be used, and when this is done it is found that the transitions are allowed. Secondly, on going from the symmetry of the undistorted molecule to a lower symmetry by a Jahn–Teller effect, one component of the active vibration that brings about the distortion becomes a totally symmetric vibration in the new symmetry. Its excitation is then allowed on ionization, and if the change in equilibrium nuclear positions is substantial, strong excitation is likely. If the change in shape is small and there are no other complications, the intensity of the line showing excitation of one quantum of the active vibration divided by the intensity of the (0–0) transition should be equal to the parameter D. The excitation of degenerate modes on ionization is hence the first indication of Jahn–Teller effects. Examples already given are the first bands in the spectrum of benzene and in that of methyl bromide.

In the benzene and methyl bromide ionizations, the Jahn–Teller distortion is small, as the (0–0) vibrational transitions are still the strongest. When the vibronic interaction is stronger, the relevant parts of the potential energy surfaces must be considered in more detail. *Figure 6.9* shows the potential energy as a function of the distortion co-ordinate, ρ, for a transition from a non-degenerate ground state to a doubly degenerate ionic state.

The distortion co-ordinate, ρ, is compounded from the two normal co-ordinates, q_1 and q_2, of the active vibration, which are perpendicular to one another. The definition of ρ is:

$$\rho = (q_1^2 + q_2^2)^{\frac{1}{2}} \qquad (6.6)$$

Now, it can be proved[2, 11] that if only linear terms are taken in the vibronic interaction, the potential energy depends only on ρ and not on the individual values of q_1 and q_2, so that the potential energy surface depicted in *Figure 6.9* is cylindrically symmetrical and must be imagined as having rotational symmetry about the central vertical axis. It is the neglected quadratic and higher terms that produce separate minima in the potential energy surface, such as the three for methyl chloride ions mentioned earlier. The potential energy surface in the upper state has its minimum at a non-zero value of ρ while the minimum in the lower state is at $\rho = 0$, but its probability density, which is shown in *Figure 6.9*, is at a non-zero value of ρ because of the cylindrical symmetry in the q_1, q_2 space. The probability density is obtained by multiplying the square of the wave-function by $2\pi\rho$, the volume of the space available at ρ. Vertical transitions take place from the position of highest probability density of the ground-state vibration as shown in *Figure 6.9*, and because of the Jahn–Teller distortion they cross the upper potential energy surface at two places. There are now two vertical transitions of different energies and the photoelectron band contour may have two separate maxima. Several instances of ^2E states showing this

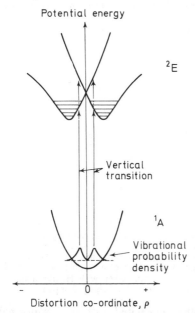

Figure 6.9. Section through the potential energy surfaces involved in a transition from a simple ground state to a ^2E ionic state undergoing a static Jahn–Teller effect

type of splitting have been found in photoelectron spectra and an example is shown in *Figure 6.10*. This is the origin of the typical Jahn–Teller band contour, that is, of bands of type 3 (Chapter 5, Section 5.4.2).

By using the semi-classical Franck–Condon approximation and harmonic oscillator potentials, an expression for the splitting, ΔE, between the two observed intensity maxima can be obtained[11] in terms of the parameter D and the frequency of the active vibration, v; the frequency is assumed to be the same in the ion as in the ground state of the molecule:

$$\Delta E = 2hvD^{\frac{1}{2}} \tag{6.7}$$

The splitting between the maxima is smaller than the Jahn–Teller stabilization energy hvD provided that D is greater than four, but for smaller values of D the splitting between the maxima is greater than the stabilization. This behaviour is confirmed by an exact calculation of the expected intensity patterns[12]. Unless vibrational structure is resolved, the value of v needed to deduce D from equation 6.7 must be obtained from the infrared or Raman spectrum with the help of a theoretical prediction of the identity of the active mode.

In discussing the splitting of the bands, it has been tacitly assumed that the active mode is the only mode to be excited strongly on ionization. This may be so, but there is no reason why totally

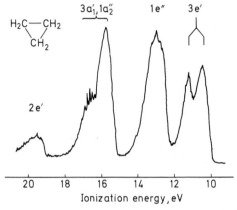

Figure 6.10. The He I photoelectron spectrum of cyclopropane showing Jahn–Teller splitting in the $3e'^{-1}$ ionization band. The $1e''^{-1}$ and $2e'^{-1}$ ionizations cause much smaller Jahn–Teller splittings, which are not apparent in the spectrum. The relative band intensities are apparently distorted by the effects of autoionization at 584 Å, since they are completely different in the spectrum taken with He II radiation[13] (From Evans et al.[14], by courtesy of Elsevier Publishing Company)

symmetric modes should not also be excited, just as in a normal transition. A noticeable static Jahn–Teller effect is likely to occur only if the degenerate orbital from which ionization takes place is strongly bonding or antibonding, and excitation of totally symmetric vibrations is therefore to be expected. The band contour for such excitation will be symmetrical with its maximum at the normal vertical ionization energy, and so tends to obscure the Jahn–Teller splitting. The vibrational structure becomes very complicated because it must still be formed from progressions of progressions; two progressions cannot simply be added together starting from the (0–0) band.

6.3.2   VIBRONIC SPLITTINGS

The other effect that one might hope to observe in photoelectron bands that show strong Jahn–Teller interactions is the splitting of the vibronic levels in the ionic state for excitation of one or more quanta of the active vibration; this would be a direct observation of the dynamic Jahn–Teller effect. The vibronic interaction splits each level in the degenerate state according to the number of quanta in the active vibration, and there are as many vibronic levels as there are symmetry species that result from multiplication of the electronic and vibrational species in the original symmetry. In an E state of a C_{3v} molecule the following species result from excitation of the first few quanta of an e active vibration:

	Vibrational species	Vibronic species
$v' = 0$	a_1	E
$v' = 1$	e	$E + A_1 + A_2$
$v' = 2$	$e \times e$	$E + E + A_1 + A_2$

As usual, this group theory indicates only the number and names of the states that may be present, and nothing about their spacings. In the approximation of a cylindrically symmetrical potential energy surface (*Figure 6.9*), it has been shown for this particular instance of an E state in C_{3v} that the A_1 and A_2 levels are degenerate[15], and a simple equation has been given for the energies[11]. The form of the vibrational structure that might actually be observed in the spectra also depends on the selection rules, as not all the new vibronic levels can be reached in direct photoionization. In ordinary optical transitions, at least for small or moderate vibronic coupling, only one vibronic component of each level can be reached in absorption starting from the vibrationless ground state[2]. The splittings cannot be observed in absorption unless more than one quantum of an active vibration is already excited in the molecule before the optical transition occurs, that is, in hot bands. The way in which this conclusion must be modified for strong vibronic coupling or for photoionization, where the outgoing electron wave can carry different amounts of angular momentum, is not yet known. It is possible that some splittings may be observable, but their absence must be taken as most likely *a priori*. If any splittings were found, the angular distribution of the photoelectrons might be different for the different vibronic components.

6.3.3 TRIPLY DEGENERATE STATES

The discussion up to this point has concentrated on doubly degenerate states, because for these states the Jahn–Teller effects are relatively easy to explain. In molecules of tetrahedral or octahedral symmetry there are triply degenerate as well as doubly degenerate orbitals, and also triply and doubly degenerate vibrations. As far as the doubly degenerate (E) states are concerned, the preceding discussion is fully applicable, but in triply degenerate (T) states the situation as regards band shape and vibronic structure is more complicated. A T state is susceptible to distortion via either the e or the t vibrational modes or a mixture of the two. Application of the semiclassical Franck–Condon approximation leads to the conclusion that ionization to a ^2T state in which only an e mode is active should give a single band in the spectrum, while activity of a t mode should give a band with three maxima. According to the Franck–Condon approximation, the central maximum coincides with the energy of the upper state at the symmetrical position, and the two outer maxima are symmetrical about it; an analytical equation for the

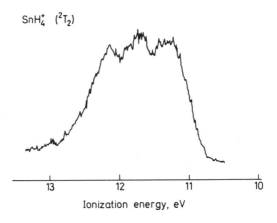

SnH$_4^+$ ($^2T_2$)

Ionization energy, eV

Figure 6.11. $^2T_2$ band in the photoelectron spectrum of stannane, showing the three intensity maxima produced by the Jahn–Teller effect. (From Potts and Price[19], by courtesy of the Council of the Royal Society)

splitting has been derived by Toyazawa and Inoue[16]. These conclusions about the band shape were confirmed by a quantum mechanical calculation by Dixon[15] for CH$_4^+$ (^2T), the case that has been studied most extensively. The photoelectron spectra of methane, silane, germane and stannane (*Figure 6.11*) do show three maxima or shoulders, in good agreement with the qualitative and also the quantitative predictions.

In methane ions formed by t_2^{-1} ionization, the most stable state is predicted to be one reached by distortion brought about by an e type vibration[17], and the predominant vibrational structure near the onset of the ^2T band does seem to be a progression in the e vibration, v_2. However, the vibrational structure is complex, conceivably including some vibronic splittings[18], and it has not yet been satisfactorily explained. The complexities of the vibrational structure might also be due to the presence of only a very small or zero barrier to inversion through the square-planar (D_{4h}) configuration in methane ions[19]. The rule that the species of the vibronic levels in the upper state are given by the product of electronic and vibrational symmetries holds for T states as well as E states, but here the splittings and selection rules are, at present, even less well understood.

6.3.4 THE MAGNITUDE OF JAHN–TELLER DISTORTIONS

If a photoelectron band is attributed to a degenerate ionic state with Jahn–Teller distortion, the semi-classical Franck–Condon approximation can be used to deduce the magnitude of the changes of bond length or bond angle that occur, as in normal bands. The energy of the ionic state in the symmetrical position is given by the point midway between the two maxima for a ^2E band, or by the central maximum for a ^2T band, and this energy takes the place of the vertical ionization potential. It is very difficult to be sure experimentally that the adiabatic ionization potential is seen in the spectrum, but equation 6.7, which relates Jahn–Teller band splitting to the stabilization energy for ^2E states, may be of use. The stabilization energy, hvD, is equivalent to the difference between the vertical and adiabatic ionization energy for a normal band.

Potts and Price[19] have made calculations of the changes in shape following t_2^{-1} ionization in the Group IV hydrides using the approximate expression 5.6. They calculated the larger HXH angles in CH_4^+, SiH_4^+, GeH_4^+ and SnH_4^+ to be 165, 150, 149 and 149 degrees, respectively, at the lowest equilibrium positions, that is, in the most stable states (all $^2B_2$ states) produced by the Jahn–Teller splitting. They also used the same method to calculate the maximum bond angles reached at different points in the bands and were able to explain the loss of vibrational structure in the SiH_4^+ and GeH_4^+ bands by the attainment of the square planar configurations of the ions in which the angles concerned reach 180 degrees. The loss of resolved vibrational structure at such a point is due to a disturbance of the vibrational levels and the halving of the frequency above the barrier of a potential energy surface with more than one minimum (see Section 6.5).

6.4   JAHN–TELLER *versus* SPIN–ORBIT EFFECTS

Just as for spin–orbit interactions, accidental degeneracy of electronic states is also a basis for Jahn–Teller effects. One interesting theoretical case of this is degeneracy of u and g states in a linear symmetrical BAB molecule, which could lead to a Jahn–Teller distortion in which one AB bond becomes longer than the other[15]. No example of this is yet known, but a similar effect might occur in an ionic state of allene[20]. The possibility of Jahn–Teller instability in accidental degeneracy must be borne in mind as another limitation on Koopmans' theorem, particularly when bands overlap.

It is now clear that most states of molecular ions produced by ionization from degenerate orbitals are susceptible to spin–orbit splittings and also to Jahn–Teller effects. The two effects are, in a sense, competing to lift the degeneracy, and details of the photoelectron spectrum depend on which effect is dominant. The strengths of the two interactions depend on different factors, spin–orbit splittings on the characteristics of the atomic orbitals involved and Jahn–Teller splittings on the bonding power of the electron removed. When one type of splitting is much stronger than the other, either normal Jahn–Teller or spin–orbit split bands are seen in the spectrum, but when they are of about the same magnitude new complications can appear that have yet another name, the Ham effect[11]. The changes expected in photoelectron bands as the splitting mechanism goes over from spin–orbit to Jahn–Teller interaction are roughly as follows.

(1) Spin–orbit ≫ Jahn–Teller. The spacings between the spin–orbit split components are larger than the quanta of the Jahn–Teller active modes and an uncomplicated spin–orbit effect is observed, as with CH_3I^+ and CI_3H^+.

(2) Spin–orbit > Jahn–Teller. If the spin–orbit splitting is comparable with the energy of the Jahn–Teller active quanta, anomalous intensity distributions may be seen owing to intensity borrowing, as in the first ionization bands in the spectra of CH_3Br and CH_3Cl.

(3) Jahn–Teller ≫ Spin–orbit. This is the region of the Ham effect, in which the same components are seen as in *(1)*, but with anomalous intensities as in *(2)* and also with shifts of the levels producing an apparent reduction of the spin–orbit splitting, possibly by a large factor. No examples are yet known from photoelectron spectra.

(4) Jahn–Teller ≫ Spin–orbit. When the Jahn–Teller effect is dominant and a static effect is produced, the spin–orbit interaction is quenched. It is not known what would be observed in a situation where both the spin–orbit interaction and Jahn–Teller effect were

very strong, except that there must still be as many bands as the original degenerate state had components.

6.5 RENNER–TELLER EFFECTS AND MULTIPLE POTENTIAL ENERGY MINIMA

Linear molecules or ions in Π, Δ or Φ electronic states undergo no Jahn–Teller distortion, but an analogue of the dynamic Jahn–Teller effect exists for them, called the Renner–Teller effect. Vibrational levels in which one or more quanta of degenerate bending vibrations are excited are split into several components by interaction of electronic and vibrational motion. In the ionization of linear molecules, however, the selection rule against excitation of bending vibrations, $\Delta v_k = 0, (\pm 2, \pm 4 \ldots)$, is not relaxed very much even for strong vibronic coupling, so this effect does not play an important role in their photoelectron spectra. The structure that one might observe in the spectrum depends not only on the vibronic splittings but also on the selection rules for excitation of the individual new vibronic levels on ionization.

Bending vibrations of linear ions are strongly excited on photo-ionization if the molecule is bent but the ion is linear, or if a large change in equilibrium angle is caused by ionization. The second ionization band in the photoelectron spectrum of water is of this type, as the ion is linear in the $^2A_1$ state produced by ionization of a $2a_1$ electron while the ground-state molecule is bent. Although the bending vibration is strongly excited, the electronic state of the ion is not degenerate, so no simple Renner–Teller effect results. Nevertheless, the second ionic state of H_2O^+ has a striking feature in that the frequency of the bending mode v_2 in this state, and also in the related Rydberg states, is only about half that in the neutral molecule. This drastic change in frequency is characteristic of a bent to linear transition because of the lifting of the inversion degeneracy. In the bent configuration, the state is degenerate in the sense that an equivalent configuration can be reached by bending the H atoms over on to the other side of the oxygen atom through the linear position. The graph of the potential energy as a function of the bending co-ordinate therefore has two equivalent minima (*Figure 6.12*) and all vibrational levels below the potential barrier are doubly degenerate on this account. Above the potential energy barrier, the previously degenerate levels are split, so twice as many vibrational levels now occupy the same energy intervals and the frequency is apparently halved. In an ionization process of a molecule going from a bent

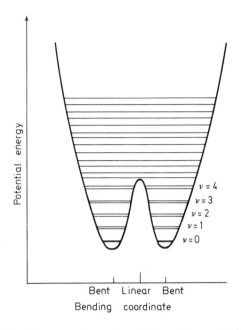

Figure 6.12. Potential energy surface with a double minimum, showing the apparent doubling of vibrational frequency above the barrier

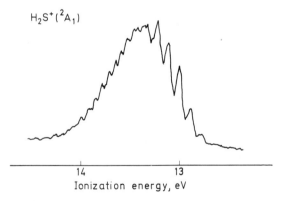

Figure 6.13. Second band in the photoelectron spectrum of hydrogen sulphide, showing the change in vibrational structure above the barrier of a double-minimum potential. (From Potts and Price[21], by courtesy of the Council of the Royal Society)

to a linear configuration or from a non-planar to a planar configuration, one therefore expects the frequency of the mode that corresponds most closely to inversion to be halved, if the force constants remain the same. Such halving of the frequency between molecule and ion has been found in the ground states of the ions PH_3^+, AsH_3^+ and SbH_3^+ as well as in the $^2A_1$ state of H_2O^+. If, in a particular state, an ion is only slightly bent, so that vibrational levels above as well as below the potential energy barrier can be reached in ionization, a change of vibrational spacings from hv' to $hv'/2$ should be seen within the spectral band. Such effects have been recognized in the second bands, which are a_1^{-1} ionizations, in the photoelectron spectra of H_2S (*Figure 6.13*), H_2Se and H_2Te^{21}.

Semi-classical Franck–Condon calculation of the change in angle from the difference between vertical and adiabatic ionization potentials confirms in each instance that the changeover takes place when the vibrational excursions are sufficient to make the ions approximately linear at the end of their motion. In earlier studies on hydrogen sulphide with lower resolution and signal-to-noise ratio, it seemed that the structure of the band simply broke off above the barrier, and this is still the apparent situation in the spectra of hydrogen selenide and hydrogen telluride. The vibrational structure above the barrier in H_2S^+ ($^2A_1$) is very complex, and only a partial analysis has been possible[22]. Once the ion attains the linear configuration, the $^2A_1$ state can interact with higher vibronic levels of the $^2B_1$ ionic ground state, because $^2A_1$ and $^2B_1$ combine to form the degenerate $^2\Pi$ state in a linear ion. This gives rise to a dynamic Renner–Teller effect and an increased complexity of the vibronic structure.

REFERENCES

1. CONDON, E. U. and SHORTLEY, G. H., *The Theory of Atomic Spectra*, Cambridge University Press, Cambridge (1935)
2. HERZBERG, G., *Electronic Spectra and Electronic Structure of Polyatomic Molecules*, Van Nostrand, Princeton, N.J. (1966)
3. WITTEL, K., *Chem. Phys. Lett.*, **15**, 555 (1972)
4. ANDERSON, C. P., MAMANTOV, G., BULL, W. E., GRIMM, F. A., CARVER, J. C. and CARLSON, T. A., *Chem. Phys. Lett.*, **12**, 137 (1971)
5. HAINK, H. J., HEILBRONNER, E., HORNUNG, V. and KLOSTER-JENSEN, E., *Helv. Chim. Acta*, **53**, 1073 (1970)
6. HEILBRONNER, E., HORNUNG, V. and KLOSTER-JENSEN, E., *Helv. Chim. Acta*, **53**, 331 (1970)
7. BRUNDLE, C. R. and JONES, G. R., *J. chem. Soc. Faraday Trans. II*, **68**, 959 (1972)
8. KING, G. H., KRISHNAMURTHY, S. S., LAPPERT, M. F. and PEDLEY, J. B., *Discuss. Faraday Soc.*, **54**, 70 (1973)

156   PHOTOELECTRON BAND STRUCTURE—II

9. BROGLI, F. and HEILBRONNER, E., *Helv. Chim. Acta,* **54**, 1423 (1971)
10. JAHN, H. A. and TELLER, E., *Proc. R. Soc., Lond.*, **A161**, 220 (1937)
11. STURGE, M. D., *Solid State Phys.*, **20**, 92 (1967)
12. LONGUET-HIGGINS, H. C., ÖPIK, U., PRYCE, M. H. L. and SACK, R. A., *Proc. R. Soc., Lond.*, **A244**, 1 (1958)
13. LINDHOLME, E., FRIDH, C. and ÅSBRINK, L., *Discuss. Faraday Soc.*, **54**, 127 (1973)
14. EVANS, S., JOACHIM, P. J., ORCHARD, A. F. and TURNER, D. W., *Int. J. Mass Spectrom. Ion Phys.*, **9**, 41 (1972)
15. ÖPIK, L. H. and PRYCE, M. H. L., *Proc. R. Soc., Lond.*, **A238**, 425 (1957)
16. TOYAZAWA, Y. and INOUE, M., *J. Phys. Soc. Japan*, **20**, 1289 (1965) and **21**, 1663 (1966)
17. DIXON, R. N., *Molec. Phys.*, **20**, 113 (1971)
18. RABALAIS, J. W., BERGMARK, T., WERME, L. O., KARLSSON, L. and SIEGBAHN, K., *Physica Scripta*, **3**, 13 (1971)
19. POTTS, A. W. and PRICE, W. C., *Proc. R. Soc., Lond.*, **A326**, 165 (1972)
20. HASELBACH, E., *Chem. Phys. Lett.*, **7**, 428 (1970)
21. POTTS, A. W. and PRICE, W. C., *Proc. R. Soc., Lond.*, **A326**, 181 (1972)
22. DIXON, R. N., DUXBURY, G., HORANI, M. and ROSTAS, J., *Molec. Phys.*, **22**, 977 (1971)

7 Dissociations of Positive Ions

7.1 INTRODUCTION

In photoelectron spectroscopy, attention is naturally focused first on the electrons, which indicate the energies of states of the molecular positive ions. The electrons alone give very little information, however, about the fates of the molecular ions after the instant of ionization, although these are also characteristic of the ionic energy states. An excited molecular ion very often dissociates into a smaller ion plus a neutral fragment, and such dissociation is the main topic of this chapter. The dissociation of excited positive ions is the basis of mass spectrometry and also has important consequences in atmospheric physics and space research. Because the ions are isolated, that is, their lifetimes are shorter than the mean time between collisions, their dissociations are examples of the simplest type of chemical rate process, unimolecular reactions. Photoelectron spectroscopy can contribute to the understanding of such elementary processes because it provides a technique by which molecular ions are produced in identified, highly excited electronic and vibrational states in known proportions. This is in contrast to electron impact ionization, the commonest source of positive ions, where neither the identities nor the proportions of the ionic states initially produced are known.

The object of examining molecular ion dissociations in detail is to answer some of the following questions. If an ion is initially in a particular quantum state and thus has a specific energy, will it dissociate? If it does, which of the energetically possible products will be formed and how fast will they be formed? How will any

157

excess energy be distributed among the products—as translational energy, or as electronic, vibrational or rotational energy of the fragments? These questions about the most elementary possible chemical reaction, a unimolecular decomposition of an isolated molecule in a defined initial quantum state, cannot be answered satisfactorily by any of the current theories of chemical kinetics.

A matter of possible importance for photoelectron spectroscopy arises from the above questions. If it should be proved that molecular ions, in particular electronic states, dissociate specifically to particular products and not just randomly* to all the products that are energetically accessible, then one should look for a relationship between the nature of the fragmentation process and that of the electronic state. A correspondence might be found between the bonding character of the electron that is missing in a particular ionic state and the bonds that are broken in fragmentation. Such a discovery, besides having important consequences for reaction rate theory, would provide a useful tool for the assignment of photoelectron spectra. Despite the apparent assumption by a number of workers that a correspondence exists, this correspondence has not yet been demonstrated experimentally. On the contrary, the success of the statistical theories of unimolecular decomposition argues rather for the conclusion that no such relationship can exist, at least for large polyatomic ions.

7.2   THE OBSERVATION OF IONIC DISSOCIATIONS

Whether or not an ion can dissociate at all is a question of energies, because for dissociation to be possible the energy of the ion must be higher than the ground-state energy of some set of possible products. Whether dissociation will take place depends on competition from the only other process that can remove the excess energy of an isolated ion without fragmentation, namely, fluorescence to a stable ionic state. The characteristic lifetime for allowed transitions that emit light in the visible or near ultraviolet spectral region is of the order of 10^{-8} s, so if a particular ionic state has an allowed transition to any stable lower state it must dissociate within 10^{-8} s if it is to do so at all. If all transitions out of the excited ionic state to stable states are forbidden, or there are no stable ionic states, then slower dissociation can occur, and conversely if a slower dissociation is observed the ionic state involved must be metastable

---

* In the sense of the statistical theories of unimolecular reactions.

in the spectroscopic sense of having no allowed transitions to lower stable states.

Whether a dissociation which does occur will be observed in a particular experiment is also a question of rates. In order for noticeable broadening of the vibrational structure in a photoelectron spectrum to occur, the time of dissociation must be very short (about 10^{-13} s), approximating to immediate breakdown. In high-resolution absorption spectroscopy, the width of a rotational line may be limited only by the Doppler effect and a broadening of 0.1 cm^{-1} may be observed, which corresponds to a lifetime of about 10^{-10} s. In mass spectrometry, on the other hand, the characteristic apparatus time is about 10^{-5} s and all dissociations that occur within this time will be detected. This encompasses almost all elementary ion decomposition processes and in addition special mass spectrometers can be built in order to study the few ion reactions with dissociation lifetimes as long as 10^{-4} to 10^{-3} s. Hence mass spectrometry is a much more sensitive test for dissociation than line broadening. The breaking-off in intensity of an emission band at the point where dissociation takes over from fluorescence is also a sensitive indication of the occurrence of dissociation, but is not very widely applicable. If all of the ions in a particular electronic state dissociate, there will be no corresponding emission band and this lack of an observed band is negative evidence, which must be used with caution. The existence of an emission band from a particular ionic state of a polyatomic molecule, on the other hand, is good evidence that dissociation from that state does not occur. The quantum yield of fluorescence is also required, however, in case fluorescence and dissociation are competing throughout the band, or the selection rules allow one set of sublevels to dissociate and a different set to fluoresce, as is known to occur in some diatomic pre-dissociations.[1]

Of the established techniques that are based on mass spectrometry, the most useful for investigating ionic dissociations are photoionization mass spectrometry[2] and charge-exchange mass spectrometry[3]. In photoionization mass spectrometry, ionization is brought about by ultraviolet light, the wavelength of which can be varied by means of a monochromator. The intensity of each ion peak in the mass spectrum is measured as a function of the photon energy, and the results are presented as a graph of ion signal per incident photon against photon energy, called a photoionization yield curve. When the photon energy reaches the threshold for the production of a new state of the molecular ion, a step should ideally appear in the yield curve either of the parent ion or of the ion formed by fragmentation. However, autoionization causes great difficulty in interpreting yield curves in terms of the production of different ionic states, and it is very seldom possible to ascribe the production of a particular fragment to dissociation from a particular ionic state.

In charge-exchange mass spectrometry, ionization of the sample, M, is brought about by transfer of an electron to an already ionized species, A^+, instead of by photon or electron impact:

$$A^+ + M \rightarrow A + M^+ \tag{7.1}$$

Provided that the ionization potential of A is greater than that of M and the particles collide with very little relative kinetic energy transfer, the following energy balance can be written:

$$I(A) = I(M) + E^*(M^+) \tag{7.2}$$

A precisely known amount of energy, the ionization energy of A, is available, so the excitation energy given to M^+ is also exactly defined. The mass spectra then show the products that are formed from M^+ as the nature of A^+, which determines the excitation energy of M^+, is varied. The variation is possible only in rather coarse steps because few ions suitable as A^+ are available, but nevertheless the information obtained from these difficult experiments is very useful. It can be brought into the form of a breakdown diagram, which shows the fraction of the ions that end up as particular product ions as a function of the internal excitation energy. If the breakdown diagram and the photoelectron spectrum of a molecule are known, the mass spectrum produced by the same radiation as the photoelectron spectrum can be predicted.

Of the methods mentioned so far, the spectroscopic methods can show which states of a molecular ion dissociate, while those based on mass spectrometry generally show the product ions that are formed but not the states of the molecular ion from which they are formed. The combination of photoelectron spectroscopy and mass spectrometry, called photoelectron–photoion coincidence spectroscopy and discussed in Section 7.7, is the only method by which the relationship between the nature of the initial ionic states and the identity of the dissociation products can be studied directly.

7.3 MODELS OF MOLECULAR AND IONIC DISSOCIATION

In considering dissociation mechanisms, one is concerned with how the excitation energy given to the molecular ion is distributed in the course of time until, and indeed after, fragmentation occurs. Three models are useful as prototypes for real dissociations, namely *direct dissociation*, *pre-dissociation*, which may be *electronic* or *vibrational*, and *internal conversion* followed by vibrational pre-dissociation. In direct dissociation, the excitation energy starts and remains in the reaction co-ordinate, but in the other mechanisms it is distributed between the available degrees of freedom in a more complicated manner.

7.3.1 DIRECT DISSOCIATION

If the state of a diatomic ion reached directly in photoionization is repulsive (unbound) or a repulsive part of a bound surface, the

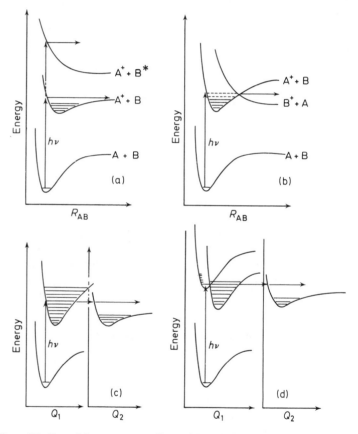

Figure 7.1. Potential energy curves illustrating some important dissociation mechanisms: (a) direct dissociation; (b) electronic pre-dissociation; (c) vibrational pre-dissociation; (d) internal conversion with vibrational pre-dissociation. In (c) and (d) the co-ordinates Q_1 and Q_2 represent different molecular motions in a polyatomic molecule

molecular ion will dissociate immediately and there can be no vibrational structure in the photoelectron spectrum (*Figure 7.1(a)*). Conversely, if it can be proved that lack of vibrational structure in a photoelectron band is due to dissociation, then the potential energy surface reached in ionization is effectively unbound even if according to a normal model of the electronic structure it is expected to be a bound state. An otherwise bound state may be effectively repulsive because of interaction with a repulsive state involving curve crossing or avoided crossing. The products will fly apart with most of the excess energy of the molecular ion above the dissociation limit as translational energy, as there is insufficient time available for it to be redistributed among internal modes before the fragments separate. The potential energy curves representing this process in *Figure 7.1* are suitable only for diatomic molecules; for polyatomic molecules, they must be imagined as two-dimensional cuts through multi-dimensional surfaces with a reaction co-ordinate as the horizontal scale. They retain their simple meanings for dissociations and spectra of polyatomic species only if this reaction co-ordinate and the vibrational motions excited by ionization are the same. It is unlikely that the change in equilibrium molecular geometry produced by ionization of a polyatomic molecule will correspond exactly and exclusively to motion in a reaction co-ordinate; direct dissociation in the strict sense can be defined as a process in which it very nearly does so.

True direct dissociation as defined above can occur only if the removal of an electron has a strong and specific influence on the bonding in the molecule. A close relationship between the identity of the products and the bonding character of the ionized electron is therefore to be sought in such instances. Unfortunately, there are as yet no clearly characterized direct dissociations of ions for which this idea can be tested. The detection of a particular product can be considered to be significant only if an alternative decomposition is energetically possible, and this is not so in the few direct dissociations of polyatomic ions that have been recognized up to now.

7.3.2   PRE-DISSOCIATION

If a molecular ionic state is definitely bound, as it has vibrational structure in the spectrum, but nevertheless dissociation takes place, the process is called pre-dissociation. Herzberg[1] has classified pre-dissociations into three ideal cases, as follows.

In *Case I*, pre-dissociation occurs by rearrangement of electronic energy; there is a non-radiative transition from the bound state to a continuous state, that is, from one potential energy surface to another. If the two electronic states have the same symmetry species, the pre-dissociation is *homogeneous*; if they are of different symmetry, it is *heterogeneous*. Electronic pre-dissociation is probably the most common mechanism by which small molecular ions formed in photoionization decompose. The way in which the excess energy is divided between the degrees of freedom of the products, including translation, depends on the details of the potential energy surfaces, especially at the configuration where the transition takes place between them. It is possible for a large fraction of the energy to appear as kinetic energy, but alternatively internal vibrational or rotational energy of the products may be favoured. Some correlations between the bonding power of an electron ionized and the identity of the products formed may be sought, as the nature of the electron ionized determines the forms of the initial ionic potential energy surface and of the initial vibrational motion. However, the existence of such correlations will also depend on several other factors, particularly the forms of surfaces that lead to different products and potentially cause the pre-dissociation, and the strengths of their interactions with the bound potential surface.

In *Case II*, pre-dissociation takes place on a single electronic energy surface by rearrangement of the vibrational energy. The excited molecule has sufficient energy to dissociate but the energy is initially in vibrational modes that do not correspond to the reaction co-ordinate for dissociation. Most thermally induced unimolecular reactions are such vibrational pre-dissociations and involve the ground electronic state only. There is very little evidence to show how common this mechanism may be among molecular ion dissociations, but it is probably less important for small ions than electronic pre-dissociation. A more distant relationship between the identity of the products and that of the electron removed must be expected in vibrational pre-dissociations than is possible in *Case I*, as the vibrational energy is removed from the mode excited on ionization into other modes, which presumably correspond more closely to different reaction co-ordinates. According to theories of unimolecular reactions, the excess energy should be distributed between all the degrees of freedom of the products, in favourable instances in a statistical manner.

Case III, in which dissociation is the result of the rotational energy which a molecule possesses in excess of a dissociation limit, will not be considered further here.

7.3.3   INTERNAL CONVERSION PLUS VIBRATIONAL PRE-DISSOCIATION

Because of its supposed prevalence, this mixed mechanism must be considered separately, although it is a sub-case of pre-dissociation. According to the quasi-equilibrium theory of mass spectra (QET), all excited molecular ions relax rapidly by conversion of their electronic excitation energy into vibrational energy of the molecular ion in its electronic ground state, after which fragmentation follows by vibrational pre-dissociation. The first step is an internal conversion from one bound electronic state to another, a process that is strictly controlled by selection rules in small molecules but thought to reach completion in 10^{-11} to 10^{-12} s in larger molecules[4]. In this mechanism, all traces of the identity of the electron originally ionized are lost and the abundances of different products should depend only on the dissociation limits for their formation and on statistical factors. The excess energy must be distributed statistically among all the internal degrees of freedom of the fragments.

Potential energy curves intended to illustrate these three mechanisms are shown in *Figure 7.1*, but it must be emphasized that the representation is very crude. At least two cuts must be made through the multi-dimensional potential energy surfaces in order to indicate vibrational pre-dissociation at all, whereas the molecules themselves have a choice from an infinite number of such cuts. Very many more sub-cases of pre-dissociation can be imagined than are represented in *Figure 7.1*, and a wider selection has been illustrated by Mulliken[5].

7.4   RATES OF ELEMENTARY PROCESSES

Which of the models of ionic dissociation best describes a given decomposition depends first on the nature of the energy surfaces, particularly whether direct dissociation is possible or not, and then on the rates of the competing elementary processes. Clues to the rates of the elementary processes are provided by the correlation rules, by selection rules and, for vibrational pre-dissociations, by statistical theories of unimolecular reactions. These last are unfortunately the main clues that are available to the rates of dissociation of most large polyatomic molecular ions because the application of correlation rules or selection rules to species with no symmetry gives little useful information.

7.4.1 CORRELATION RULES

If direct dissociation is to occur, the state of the molecular ion reached in photoionization must go over directly to a lower dissociation limit (*Figure 7.1(a)*). The symmetry species of the molecular ionic state must be identical with a species that can be obtained by combining the products; the correlation rules indicate those species of states of the molecular ion which can be obtained by combining particular products, and therefore whether direct dissociation to these products is possible or not. The correlation rules, also called Wigner–Witmer rules, are given by Herzberg (reference 6, p. 281), whose book is a most useful source of the symmetry tables needed in applying these rules; for diatomic molecules,: the rules are presented in a convenient form by Hasted[7]. In essence, a point group is chosen that corresponds to the symmetry that is conserved in the dissociation, and the appropriate tables (reference 6, p. 574) are used in order to find the species of the molecular state and the states of the separated fragments in this group. The direct products of the representations to which the states of the fragments belong (reference 6, p. 570) then give the orbital species that can result from their combination. For the spins the rule is simply

$$S = S_1 + S_2, S_1 + S_2 - 1 \ldots S_1 - S_2 \qquad (7.3)$$

The multiplicity is, as always, equal to $2S + 1$, and corresponding pairwise combination is valid for more than two fragments. An example of the use of the rules is provided by the dissociation of methyl chloride ions in their first excited state, $^2A_1$, corresponding to the continuous second band in the photoelectron spectrum:

$$CH_3Cl^+ \, (^2A_1) \rightarrow CH_3^+ \, (^1A_1') + Cl \, (^2P_u) \qquad (7.4)$$

The dissociation can be assumed to take place along the C–Cl axis, so that the whole molecular symmetry (C_{3v}) is conserved. The ground state of planar CH_3^+ goes over to $^1A_1$ in C_{3v}, while the ground-state chlorine atom gives $^2A_1 + {}^2E$. The direct product of A_1 with $A_1 + E$ is again $A_1 + E$ and the spin rule allows only doublets, so a $^2A_1$ state does arise from the combination of the products. A direct dissociation is possible and is probably the reason for the continuous nature of the second band in the spectrum. For the related fragmentation of ground state CF_4^+ ions (Chapter 5, Section 5.4.1),

$$CF_4^+ \, (^2T_1) \rightarrow CF_3^+ \, (^1A_1') + F \, (^2P_u) \qquad (7.5)$$

C_{3v} symmetry may still be conserved and direct dissociation is possible through the ^2E components because in C_{3v} symmetry $^2T_1$ (of T_d) goes over to $^2A_2 + {}^2E$.

In the photoelectron spectrum of hydrogen sulphide, the second band shows an apparent breaking-off of vibrational structure at 13.3 eV (*Figure 6.13*), which happens to coincide with the dissociation limit for the process

$$H_2S^+ \rightarrow S^+ \, ({}^4S_u) + H_2 \, ({}^1\Sigma_g^+) \qquad (7.6)$$

This coincidence has given rise to suggestions that the loss of vibrational structure might be due to rapid dissociation, but as

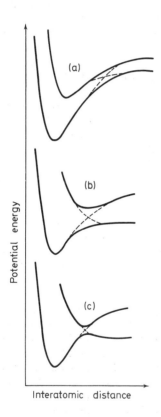

Figure 7.2. Potential energy curves for a diatomic molecule illustrating the non-crossing rule. The dashed lines indicate the unperturbed curves in each instance and the full lines the effective potential energy curves: (a) interaction between two bound states; (b) and (c) interaction between a bound state and a repulsive state, very strong and weak, respectively

the states of H_2S^+ reached in photoionization are all doublets a direct dissociation is forbidden by the spin rule. It is now known that the change in vibrational structure at 13.3 eV has a different origin, as described in Section 6.5.

The correlation rules, as with all considerations that are based purely on symmetry, indicate only the combinations which are possible and those which are impossible. In order to establish whether a particular state of the products of a dissociation goes over directly to a particular molecular state, further information is required. The most useful guide is the *non-crossing rule*, which is also based on symmetry. This rule states that in a potential energy diagram for a diatomic molecule, no curves of the same symmetry species can cross, but rather they avoid each other, thus producing two new curves. The rule is based on the idea that there is always an interaction between two states of the same symmetry species, although they may be non-interacting states in an approximate model of the system. The interaction is represented by a matrix element, $< \Psi_1 | \hat{P} | \Psi_2 >$, which must be non-zero if the operator $\hat{P}$ contains all the terms that are omitted in the approximate model. From the rule, it follows that the lowest state of a given symmetry in a diatomic molecule must go over either directly or by an avoided crossing to the lowest state of the same symmetry obtainable by combining dissociation products. Furthermore, higher states of the same symmetry must likewise go over to the higher dissociation limits of correct symmetry in their natural energetic order. The shape of the potential energy curves produced, however, depends very much on whether the interaction involved is weak or strong, as illustrated in *Figure 7.2*. Even two electronic states of exactly the same symmetry species may interact only weakly if, for instance, they differ in their electron configurations. If the interaction is weak, the rate at which the system can pass from one of the original states to the other is low, and the lifetime of a molecule in an initial bound state may be long.

For polyatomic molecules, the non-crossing rule can be considered to hold for many practical purposes, although the theory has to be modified so as to allow for the occurrence of a special form of crossings called conical intersections[6].

7.4.2 RATES OF RADIATIONLESS TRANSITIONS

Both pre-dissociation and internal conversion are examples of radiationless transitions in which the system passes from one electronic state to another at the same energy. The rate of such a process depends on two factors: the strength of the interaction between the electronic states and the agreement between the nuclear positions in the two states at that energy. The rate depends on the magnitude of a matrix element, which may be separable into an electronic part and a vibrational overlap integral. The two parts cannot always be separated because the interaction between the states may be brought about by coupling between nuclear and electronic motion. Deviations from the Born–Oppenheimer approximation must then be taken into account and potential energy curves are an incomplete model for dynamic processes. Nevertheless, it is helpful to discuss pre-dissociation in terms of deviations from the classical model of separate potential energy surfaces. If the lifetime of the initial state is τ, there may be many accessible final states that are isoenergetic within the energy uncertainty $\hbar\tau^{-1}$ and the lifetime then depends also on the density of final states, $\rho(E_f)$. These ideas are combined by the 'golden rule' of perturbation theory into an equation for the lifetime:

$$\tau^{-1} = 2\pi\hbar^{-1} P^2 \rho(E_f). \tag{7.7}$$

Here P is the matrix element of the perturbation, $P = <\Psi_f | \hat{P} | \Psi_i >$, which may be separable, as an approximation, into an electronic part and a vibrational overlap integral:

$$P \approx <\psi_f | \hat{P} | \psi_i > \; <\chi_f | \chi_i > \tag{7.8}$$

The main effect of the vibrational terms is that radiationless transitions are likely to occur only if the potential energy surfaces cross, or at least come very close to each other. Other factors being equal, the flatter the surfaces are at the crossing point, that is, the less acute is the angle at which they cross, the more extended the vibrational wave-functions are and the greater will be the transition probability. Transitions can occur not only at the crossing point but also above or below it, because of the extension of the wave-functions and the breakdown of the Born–Oppenheimer approximation. The rate will be greatest at the crossing point and will fall off on both sides. Above the crossing point, this situation can be related classically to the velocity with which the molecules move through the critical configuration; below the crossing point, the obvious analogy is with quantum mechanical tunnelling. The fall-off will be less rapid if the curves cross at an acute angle, as they then remain close together over a greater energy range. Experimental effects of these vibrational factors have been observed in the electronic spectra of pre-dissociating diatomic molecules[8].

In order to calculate the strength of an electronic interaction theoretically, the initial and final wave-functions would be required, and also a precise description of the perturbation; none of these are normally available, but clues to the strengths of interactions are given by the selection rules for perturbations, originally derived by Kronig[9]. Although the language of perturbation theory is used, no real perturbation of the molecule exists for pre-dissociation or intersystem crossing as it does, for instance, in collision processes. Instead, the perturbation operator in equation 7.8 represents terms that have been omitted in the .original approximate model of the system. The most important of these terms are interactions between electronic and nuclear motion, and spin–orbit coupling. As these terms are internal to the molecule, the operator $\hat{P}$ must be totally symmetric in the point group of the molecule and the general condition for the interaction to be non-zero is simply

$$\Gamma_f^{tot.} = \Gamma_i^{tot.} \tag{7.9}$$

where $\Gamma_f^{tot.}$, $\Gamma_i^{tot.}$ are the irreducible representations to which the total wave-functions for the final and initial states belong. The selection rules arise from translating this, together with the conservation laws that are valid for an isolated molecule, into the language of spectroscopic quantum numbers.

There are three general and rigorous selection rules, which can be written as

$$\Delta J = 0, \; + \leftrightarrow -, \; s \leftrightarrow a \tag{7.10}$$

where J is the total angular momentum, $+$ and $-$ show symmetry of the total wave-functions with respect to inversion (parity), and s and a show the symmetry with respect to exchanging identical nuclei. For internal conversions and pre-dissociations of diatomic and linear molecules for which the linear configuration is retained, these rules have important consequences. Conversion between Σ^+ and Σ^- is strictly forbidden for singlet states and becomes only weakly allowed for higher multiplets, and if a Π state would go over to a Σ state only half of the sub-levels can pre-dissociate. For linear molecules with equivalent identical nuclei, there are further restrictions imposed by the $s \leftrightarrow a$ condition, which lead to the rule for all centro-symmetric molecules that $g \leftrightarrow u$, the opposite of the rule for optical transitions. For further details of the application of the rules to linear molecules the reader is again referred to Herzberg's book[6]. For non-linear polyatomic molecules, the strict selection rules still apply, with the necessary modifications to the symmetry groups of the molecules, but they impose no practical restrictions on pre-dissociation; in real cases they can always be fulfilled.

In addition to the rigorous selection rules, there are approximate selection rules that would hold if the approximate model of the systems were correct. The approximate rules are much more restrictive than the strict rules and must be broken in most pre-dissociations. An interaction allowed by all the rules means identical symmetry species for the bound and unbound states, which will lead to a. strong interaction if the potential energy surfaces are close together.

The approximate rules and the perturbations that allow them to be broken[10] are shown in *Table 7.1*. In addition to these approximate selection rules, the strict selection rules always apply, in particular $\Delta J = 0$. In transitions brought about by spin–orbit or rotational–electronic interaction, the product of the initial and final electronic wave-functions must transform as a rotation and the rule $g \leftrightarrow u$ holds strictly.

Table 7.1   APPROXIMATE SELECTION RULES FOR RADIATIONLESS TRANSITIONS

The entries in each column show how the approximate selection rule may be broken; a zero indicates that the particular perturbation has no effect. Rotational–electronic interaction makes possible transitions with $\Delta\Lambda = \pm 1$ and/or $\Delta\Omega = \pm 1$ in the linear case, and transition between states differing by the species of a rotation in the non-linear and general case.

Rule	Perturbations		
	Spin–orbit	Rotational–electronic	Vibrational–electronic
Diatomic and linear molecules that pre-dissociate linearly			
$\Delta S = 0$	± 1	0	
$\Delta\Lambda = 0$	$\pm 1^*$	$\pm 1^*$	
$\Delta\Omega = 0$	0	$\pm 1^*$	
Non-linear molecules			
$\Delta S = 0$	± 1	0	0
$\Gamma_f^{el.} = \Gamma_i^{el.}$	Broken	$\times \Gamma^{rot.}$	$\times \Gamma^{vib.}$

* In these cases the electron configurations of the interacting states must differ by not more than one spin–orbital.

The most important of the approximate selection rules in practice is the spin selection rule $\Delta S = 0$, which can be broken only through spin–orbit coupling. For molecules that are made up from atoms of first row elements (Li to F), transitions that break this rule are likely to be 10^{-3}–10^{-5} times slower than would otherwise be the case. If a 'normal' pre-dissociation occurs in 10^{-11} s, a spin-forbidden pre-dissociation for compounds of first row elements may be just competitive with radiation, or even slower.

In order to apply the selection rules to a pre-dissociation, it is necessary to know or guess the species of the unbound state that cause it. In order to find the possible species, the states of the assumed products are combined according to the correlation rules as before, and the most significant results that can then emerge are either that for some combination of the products the pre-dissociation is fully allowed, or that it is always spin forbidden. The symmetries do not indicate the full situation, however, because even if two states have exactly the same symmetry the interaction between them may be weak if they differ in their electron configurations. An example of this is given later in the chapter.

The ions that are produced by ionizing any closed-shell molecule have an odd number of electrons, and if an ion so formed breaks into two fragments, one has an odd and the other an even number of electrons. Odd-electron species have doublet or quartet states, while even-electron fragments have singlet or triplet states; higher multiplicities are rarely encountered among molecular species. The only spin-forbidden combination in the decomposition of a doublet molecular ion into two fragments is quartet + singlet. The most important fragments that have quartet ground states are the atoms of Group V (N, P) and the isoelectronic ions of Group VI (O^+, S^+), and it is only when these species are involved that spin-forbidden pre-dissociations are likely to be encountered. Examples are the pre-dissociation from doublet states to the lowest energy products in CO_2^+, COS^+, CS_2^+ and N_2O^+. In the fragmentation of larger polyatomic molecular ions where atomic fragments are not involved, spin-forbidden processes are seldom important.

7.4.3 INTERNAL CONVERSIONS

The rate of an internal conversion depends on exactly the same factors as that of a pre-dissociation, that is, the electronic transition probability, vibrational overlap integral and the density of final states. However, as the transition is between two bound states, the restrictions imposed by selection rules are more severe, particularly for small molecules[11]. Unfortunately, there is little clear experimental evidence about the extent to which internal conversions actually occur, and this is a current problem both in photochemistry and in the theory of unimolecular reactions. The QET theory of mass spectra assumes that all excited states of molecular ions relax rapidly to the ground state by internal conversion before dissociation. This assumption implies that all primary fragmentations must be in competition with each other and that two fragmentations with different energy requirements cannot proceed at similar rates so as to give ions of comparable intensity in the mass spectrum, whatever the original distribution of excitation energy. Some measurements on the fragmentation of small molecules that contradict this requirement have been interpreted in terms of isolated electronic states, which are not interconverted[12]. On the other hand, the QET theory has been successful in interpreting the mass spectra of large molecules, both qualitatively and quantitatively, and this success may be evidence in favour of its original assumptions. Proponents of the theory claim at present that it should be valid for molecules with more than, say, five or six atoms, for which the density of states is high enough, and the restrictions due to selection rules are weak enough, to allow easy internal conversions.

Photochemical evidence concerning internal conversions is obtained from a comparison of the absolute quantum yields of fluorescence, phosphorescence and intersystem crossing for molecules in solution or rigid glasses. There is unfortunately little information available for gases. One would expect that the continual collisions with solvent molecules in liquids would provide sufficient perturbations to make radiationless transitions universal despite any selection rules or unfavourable overlap factors, but this is not apparently so. For aromatic molecules in solution at room temperature, the conclusion that internal conversions from the first excited singlet state to the ground state singlet state play a significant role can be drawn only tentatively[8]. In some instances there is definitely no such internal conversion as the fluorescence yields are unity; examples are 1,10-dimethylanthracene and p-terphenyl in light petroleum. For such large molecules, an advocate of the QET theory would not hesitate to assume that complete internal conversion among all ionic states occurs within 10^{-12} s. On the other hand, the higher excited singlet states of aromatic molecules are effectively converted to the lowest excited singlet state by internal conversion in much less than 10^{-8} s because, with the exception of azulene and some of its derivatives, all molecules that fluoresce in solution do so from their lowest singlet state, whatever the exciting wavelength.

The state of knowledge of internal conversions is at present unsatisfactory and it is hoped that this situation will soon be improved in both the experimental and theoretical aspects. Samson et al.[13] recently presented evidence for internal conversion between two excited states of the CO_2^+ molecular ion, but it is not yet known how this transition occurs.

7.4.4   VIBRATIONAL PRE-DISSOCIATION

In vibrational pre-dissociation, which occurs on a single electronic surface, the rate of reaction depends strongly on the amount of excess energy available above the threshold value. The excitation energy is assumed to move around within the molecule from one vibrational mode to another until sufficient energy to cause dissociation accumulates in the reaction co-ordinate. If the molecule is large enough for statistical theory to apply, the rate constant is given to a first approximation by

$$k = k_0 \left(\frac{E-E_0}{E}\right)^{N-1} = k_0 \left(1 - \frac{E_0}{E}\right)^{N-1} \tag{7.11}$$

where E is the energy available, E_0 the threshold energy, N is the number of vibrational modes in the molecular ion and k_0 is of the order of a vibration frequency, say 10^{13} s^{-1}, and is characteristic of the particular dissociation and the form of the activated complex. Because the factor in parentheses is necessarily less than unity, the larger the molecule the slower is its dissociation for a given excess energy above the threshold. Near the threshold there is a minimum reaction rate for molecular ions that possess only one quantum of vibrational energy above the threshold. For triatomic molecules, this minimum rate is about 10^9 s^{-1}; for larger molecules, it rapidly becomes less, being about 10^5 s^{-1} for a penta-atomic molecule. Equation 7.11 is, however, only a first approximation as it is derived not only on the assumption of statistical equilibrium, but also of classical harmonic oscillators, all of which have the same frequency. For details of the more precise forms of equation 7.11 the reader is referred to specialist publications[12].

Only in very small molecules or for excitation energies far above the threshold can dissociation by this mechanism proceed fast enough to cause broadening of vibrational lines in a photoelectron spectrum. Broadening immediately above the threshold is definitely not to be expected, indeed it is unusual, without the intervention of internal conversion, for high enough vibrational levels of any single electronic state to be reached for pure vibrational pre-dissociation to occur. A conceptual difficulty with this idea occurs, however, in deciding what constitutes a single electronic state. Because of strong interactions, whether described as avoided crossings or perturbations that lead to radiationless transitions, the effective potential energy surface from the point of view of a dissociating molecule may be a single one, although it includes several electronic states. This situation is possibly common for large polyatomic ions. Whether a real pre-dissociation is vibrational or electronic might be decided in retrospect according to whether the rate-determining factor is reorganization of the vibrational or the electronic energy, but even this does not necessarily provide an unambiguous definition.

7.5 DISSOCIATION LIMITS

A knowledge of the dissociation limits for different fragmentation pathways is indispensable for an interpretation of molecular ion breakdown. Unfortunately, reliable information on these limits is scarce except for a few small molecules. One must distinguish between the dissociation limit, which is a theoretical quantity, the true energy difference between the products and reactants in their ground states, and the experimental approximations to it, which are the appearance potentials. It is also important to notice the distinction between dissociation *limits*, as discussed here and quoted in later examples, and dissociation *energies*. The dissociation limits are referred to the ground state of the neutral molecule as energy zero, and so for formation of ionic products they are usually greater than 10 eV. Dissociation energies are defined for both molecules and ions in particular electronic states, and are referred to the ground vibrational level of the state concerned; they are almost always less than 5 eV.

Dissociation limits can be estimated in two ways. If data are available, the dissociation limit E_0 (X) for the formation of fragment X can be estimated thermochemically by using Hess' Law:

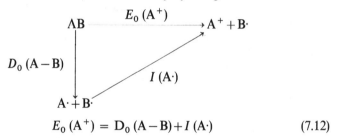

$$E_0 (A^+) = D_0 (A-B) + I (A\cdot) \qquad (7.12)$$

In this example, the data required are the dissociation energy of the ground-state molecule into two neutral fragments, $D_0 (A-B)$, and the ionization potential of one of them, $I (A\cdot)$; the sum of these two energies is the dissociation limit. Alternatively, the dissociation limit can be expressed in terms of the standard heats of formation, ΔH^0_{f0}:

$$E_0 (A^+) = \Delta H^0_{f0} (A^+) + \Delta H^0_{f0} (B\cdot) - \Delta H^0_{f0} (AB) \qquad (7.13)$$

$$= \Delta H^0_{f0} (A\cdot) + \Delta H^0_{f0} (B\cdot) + I (A\cdot) - \Delta H^0_{f0} (AB) \qquad (7.14)$$

The thermochemical quantities should refer to 0 °K rather than to room temperature, although the differences are usually small. For a few small molecules, dissociation energies, or the equivalent standard heats of formation, are known reliably from many direct and indirect measurements and the ionization energies of the fragments are known from spectroscopy, photoionization or photoelectron spectroscopy. These are the favourable instances in which the dissociation limits can be calculated within an uncertainty of about ± 0.05 eV. For the great majority of fragmentations of larger molecules, such precision is not possible, mainly because accurate heats of formation or ionization potentials of the radicals are not available. Rough estimates can be made by using thermochemical quantities derived mainly from electron-impact appearance potentials as estimates of dissociation limits. Best values of the heats of formation of many positive ions have been tabulated[14], and are weighted means of the most reliable direct or indirect measurements of these quantities. For the neutral species, recourse can be had to standard thermochemical tables[15]. The nature of the reaction products must be known or assumed before a dissociation limit can be calculated in this way. Consider the decomposition of n-hexane to give an ion of mass 57:

$$n\text{-}C_6H_{14}^+ \rightarrow C_4H_9^+ + |C_2H_5| \tag{7.15}$$

Different limits can be calculated according to whether the ionic product is n-, sec-, iso- or t-$C_4H_9^+$ and whether the neutral products are C_2H_5 radicals or $CH_3 + CH_2$ radicals, to name just a few of the more likely possibilities. In order to examine dissociation processes in detail, one must at first consider less complex examples.

The second method of estimating dissociation limits is to assume that they are equal to the appearance potentials of the fragment ions measured by electron impact or, much preferably, by photoionization mass spectrometry. This approximation is to be avoided if possible because experience shows that even appearance potentials obtained by photoionization can differ from the dissociation limits by several tenths of an electronvolt. One of the reasons is that there may be no state of the molecular ion that can be attained in direct photoionization at the threshold. The photoelectron spectra of most small molecules contain wide spaces between the ionization bands, and any dissociation limit that falls in such a space is difficult to detect. The probability of a direct transition at the threshold is very small there, and often the onset of the next allowed state of the molecular ion is measured as the appearance potential. This particular difficulty is reduced in electron impact work as many more ionic states are available, because the selection rules for electron impact ionization are less restrictive than those for photoionization; unfortunately, there are other problems. Until very recently, the energy resolution that could be attained in electron impact ionization was so low (about 1 eV) that accuracies* much better than ± 0.5 eV could not be claimed, despite the high precisions that were often obtained.

7.6   DISSOCIATIONS SEEN IN PHOTOELECTRON SPECTRA

Figure 7.3 shows photoelectron bands from the spectra of the molecules HF, HBr and HCN, all of which contain line broadening definitely due to rapid dissociation. The processes involved can now be discussed in terms of the mechanisms of direct dissociation and pre-dissociation considered in the previous sections.

HF^+

The first excited state of HF^+ is a $^2\Sigma^+$ state that arises from the ionization of an electron from the H–F bonding σ orbital. The second band in the photoelectron spectrum[16] shows vibrational structure from the onset at 19.09 eV up to 19.43 eV, where it breaks off to yield to a continuum. The lowest dissociation limit is for the production of H^+ (^1S_g) + F (^2P_u) atoms in their ground states, and calculation and experiment agree that it lies at 19.44 eV. The fact that the vibrational structure disappears exactly at the dissociation limit shows that this is a direct dissociation—that is, ionization above the limit goes to an unbound part of the upper potential energy surface. The products combine according to the correlation rules to give a $^2\Sigma^+$ state and a $^2\Pi$ state. The $^2\Pi$ combination correlates directly with the ground state of HF^+, which is X $^2\Pi$, while the $^2\Sigma^+$ combination goes over directly to the A $^2\Sigma^+$ state of HF^+, which dissociates. This is confirmed by correlation of the individual orbitals and by a detailed numerical calculation[17], and so the dissociation of HF^+(A $^2\Sigma^+$) ions is definitely simple and direct, as depicted in *Figure 7.1(a)*.

HBr^+

The first excited state of HBr^+ is a $^2\Sigma^+$ state that corresponds to ionization of an H–Br bonding σ electron. The lowest dissociation

---

* Accuracy is used here and elsewhere in this book in the sense of the deviation of a measurement from the true quantity aimed at. Precision, on the other hand, refers only to the reproducibility of a particular measurement.

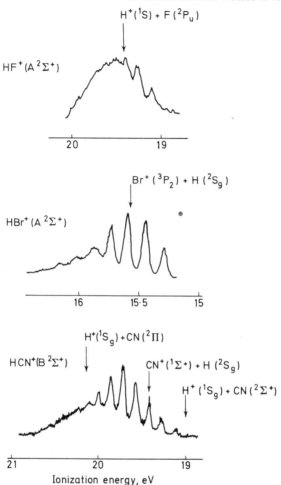

Figure 7.3. Three photoelectron bands which show broadening due to rapid dissociation. The arrows give the positions of dissociation limits for formation of the products named. (HF from Brundle[16], by courtesy of the North Holland Publishing Company; HBr from Schneider and Smith[18], by courtesy of the North Holland Publishing Company; HCN from Baker and Turner[20], by courtesy of the Council of the Royal Society)

limit is calculated to be 15.59 eV and represents the formation of the products H (^2S_g) + Br$^+$ $(^2P_2)$. This dissociation limit is well below the onset of observable broadening in the photoelectron spectrum[18], which first sets in for the $v' = 4$ line of the $^2\Sigma^+$ band at 15.9 eV. The emission spectrum of HBr$^+$ (A $^2\Sigma^+$) ions produced by photoionization at 584 Å has been measured [18, 19], and the only transitions found in it are from the $v' = 0$ and $v' = 1$ levels of HBr^+, both of which are below the dissociation limit. Hence all HBr^+ ions that are produced with energies above the dissociation limit dissociate faster than they can radiate, thus in less than about 10^{-9} s, but only at the vibrational level $v' = 4$ does the dissociation rate become fast enough to cause broadening in the photoelectron spectrum. This is clear evidence that the process is a pre-dissociation, probably involving a curve crossing at or a little above the $v' = 4$ level. In connection with the study of the emission bands, it has been shown[19] that the pre-dissociation is caused by a $^4\Pi$ state that is obtained by combining the products. The other states produced by combining them are $^{2,4}\Sigma^-$ and $^2\Pi$, so that no homogeneous dissociation is possible. The interaction between the $A\,^2\Sigma^+$ state and the $^4\Pi$ state of HBr^+ must be brought about by spin–orbit coupling, which is strong for the bromine atom.

HCN^+

As the cyanide radical can be considered to be a pseudo-halogen, one might expect similarities between the dissociation of HCN^+ and the hydrogen halide ions. The electronic state involved is again a $^2\Sigma^+$ state $(B\,^2\Sigma^+)$, derived by removing an H–CN bonding σ electron from the neutral molecule, and in agreement with this the vibration excited is the H–CN stretching vibration, v_1. The second photoelectron band contains a resolved progression in this vibration[20] with a frequency markedly less than that in the neutral molecule and showing definite convergence at higher vibrational quantum numbers. The vibrational structure seems to disappear and submerge into a continuum at about 20.3 eV.

In a photoionization investigation of hydrogen cyanide, Berkowitz et al.[21] measured the appearance potentials of H^+ and CN^+ ions, which in this instance are thought to be equal to the true dissociation limits:

$$HCN \rightarrow H^+ \,(^1S_g) + CN \,(^2\Sigma^+) \quad 19.00\,\text{eV} \quad ^2\Sigma^+ \quad (7.16)$$

$$HCN \rightarrow H \,(^2S_g) + CN^+ \,(^1\Sigma^+?) \quad 19.43\,\text{eV} \quad ^2\Sigma^+ \quad (7.17)$$

$$HCN \rightarrow H^+ \,(^1S_g) + CN \,(^2\Pi) \quad 20.15\,\text{eV} \quad ^2\Pi \quad (7.18)$$

The dissociation limit for the last process follows from the electronic spectrum of the CN radical[22] and the symmetries given on the right are those obtained by combining the products on the assumption that a linear configuration is retained in all of the dissociations. No information on the excited states of CN^+ is available and even the symmetry species of the ground state is not completely certain.

If the dissociation limits and symmetries given above are correct and no other low-lying states of CN^+ are involved, a possible explanation of the form of the second photoelectron band of HCN is as follows. The first and third dissociation limits (7.16 and 7.18) correlate with the two lowest states of HCN^+, namely with the A $^2\Pi$ state and with the X $^2\Sigma^+$ state obtained by removing a nitrogen lone-pair electron from the molecule. Both of these states lie near 13.6 eV and form the first band in the spectrum, far below the dissociation limits. For the HCN^+ (B $^2\Sigma^+$) ions to decompose to the first limit, they would first have to undergo an internal conversion process to the lower $^2\Sigma^+$ state. Although this internal conversion is homogeneous, the energy separation is so large that the Franck–Condon factors must be very unfavourable and lead to a relatively slow pre-dissociation, consistent with the resolved vibrational fine structure. Above 19.43 eV, a new homogeneous dissociation to $H + CN^+$ becomes possible, and it is not obvious why the B $^2\Sigma^+$ state of HCN^+ does not correlate directly with the CN^+ limit to give a continuum in the spectrum immediately above 19.43 eV. The reason is to be sought in the correlation of individual orbitals, which is valid if the states of the species concerned are well represented by single electron configurations[17], that is, if the electron correlation is unimportant. The electronic structures of the CN radical in its ground state and relevant excited states are[22]

$$2s\sigma_1^2 \, 2s\sigma_2^2 \, 2p\pi^4 \, 2p\sigma_N \ldots X \,^2\Sigma^+ \qquad (7.19)$$

$$CN \qquad 2s\sigma_1^2 \, 2s\sigma_2^2 \, 2p\pi^3 \, 2p\sigma_N^2 \ldots A \,^2\Pi \qquad (7.20)$$

$$2s\sigma_1^2 \, 2s\sigma_2 \, 2p\pi^4 \, 2p\sigma_N^2 \ldots B \,^2\Sigma^+ \qquad (7.21)$$

For the HCN^+ molecular ions, the electronic structures are

$$2s\sigma_1^2 \, 2s\sigma_{CH}^2 \, 2p\pi^4 \, 2p\sigma_N \ldots X \,^2\Sigma^+ \qquad (7.22)$$

$$HCN^+ \qquad 2s\sigma_1^2 \, 2s\sigma_{CH}^2 \, 2p\pi^3 \, 2p\sigma_N^2 \ldots A \,^2\Pi \qquad (7.23)$$

$$2s\sigma_1^2 \, 2s\sigma_{CH} \, 2p\pi^4 \, 2p\sigma_N^2 \ldots B \,^2\Sigma^+ \qquad (7.24)$$

The individual orbital correlations are clearly as follows:

$$2p\sigma_N \, (HCN) \rightarrow 2p\sigma_N \, (CN) \qquad (7.25)$$

$$2p\pi \, (HCN) \rightarrow 2p\pi \, (CN) \qquad (7.26)$$

$$2s\sigma_{CH} (HCN) \rightarrow 2s\sigma_2 (CN) \qquad (7.27)$$

There is no occupied orbital of HCN or HCN$^+$ that correlates with hydrogen 1s; the orbital which does so is the antibonding counterpart of $2p\sigma_N$, which can be called $2p\sigma^*$. These orbital correlations show that the three electronic states of HCN$^+$ reached in photoionization go over directly to the products H$^+$ + CN, with the CN radical in the three electronic states mentioned above. The state of HCN$^+$ to which the combination H + CN$^+$ leads directly is a highly excited one in which one electron has been promoted to the $2p\sigma^*$ orbital. Because the symmetries are the same, however, the $^2\Sigma^+$ state from the products H + CN$^+$ and the B $^2\Sigma^+$ state of HCN$^+$ will probably interact and cause a hump in the effective

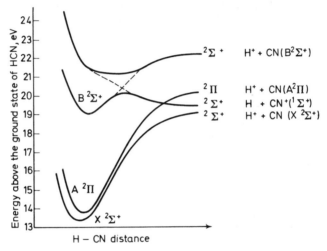

Figure 7.4. Potential energy curves for HCN$^+$ proposed as an explanation for the observed broadening in the B $^2\Sigma^+$ band

potential energy surface, as illustrated in *Figure 7.4*. Rapid dissociation of the HCN$^+$ ions to CN$^+$ will not occur at the dissociation limit, but at a higher energy characteristic of the hump, the avoided crossing. This explanation is still tentative, however, and further experimental investigation is required, perhaps by photoelectron–photoion coincidence spectroscopy.

7.7 PHOTOELECTRON–PHOTOION COINCIDENCE SPECTROSCOPY

The purpose of photoelectron–photoion coincidence measurements is to identify both of the charged particles, electron and ion, that are produced in single photoionization events. The energy of the electron shows exactly which excited state of the molecular ion has been produced, while the mass and kinetic energy of the ion show the decomposition pathway which molecular ions in that excited state follow. Two types of experiment can be carried out by using the coincidence method. First, electrons of a single energy can be chosen and the mass spectrum of ions in coincidence examined. The energy balance in photoionization makes this an investigation of the decomposition of molecular ions from known initial states and, as in normal mass spectrometry, not only the identity but also any kinetic energies of the ions produced can be measured. Secondly, particular ions in the mass spectrum can be chosen and the spectrum of photoelectrons in coincidence with them can be found. This method gives a spectrum of the excited states of the molecular ion that lead to the production of the particular ion chosen, and it is equivalent to a determination of the breakdown diagram (Section 7.2). This method of determining breakdown diagrams is free from several of the limitations of charge-exchange mass spectrometry, but has the disadvantage that nothing can be measured in empty regions of the photoelectron spectrum where no direct ionization occurs.

7.7.1   EXPERIMENTAL

The photoelectron–photoion coincidence technique requires the use of a photoelectron spectrometer and a mass spectrometer, both of which sample the same ionization events. In order to bring the ions into the mass spectrometer, an electric field is needed in the ionization region, and the strength of this field is an important experimental parameter. A schematic diagram of a coincidence spectrometer is shown in *Figure 7.5*, notional signal wave-forms being used in order to illustrate the principles.

Both electrons and ions must be detected as single pulses, and after a delay of the electron signals to compensate for the flight times of ions through the mass spectrometer the pulses are brought together in a coincidence gate. An output from this gate occurs if its two inputs are simultaneous, and if the delay is correct this should be true only of pulses from the two particles produced in single ionization events. Experimental difficulties arise from the fact that even when the delay does not match the flight time of the ions, some coincidences are still detected because of the random distribution of ion arrival times. These are accidental coincidences and their contribution must be calculated or measured and subtracted from the total signal.

All ions of the same mass would not arrive at the ion detector simultaneously, even if they were all produced at the same instant by photoionization. Part of the experimental spread in the flight times of ions is instrumental, but most of it is due to the initial velocities that the ions possess immediately upon ionization. An ion which starts out with an initial velocity towards the mass spectrometer

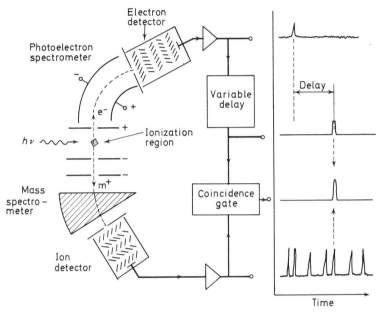

Figure 7.5. Simplified scheme of a photoelectron–photoion coincidence spectrometer. The peaks in the idealized wave-forms on the right indicate detection of single electrons and single ions; there are generally many more ions than electrons because the collection efficiency of mass spectrometers is higher than that of photoelectron spectrometers

will clearly arrive there sooner than one which starts out in the opposite direction and must first be stopped and then sent in the correct direction by the electric field. This has the effect that a curve showing the rate of true coincidences plotted against the delay time (a time-of-flight coincidence spectrum) is characteristic of the initial velocity distribution of the ions. One source of initial velocities in the photoions is the thermal gas kinetic motion of the molecules before ionization; although these velocities are small, they cause an easily observable spread of the flight times of the ions measured in coincidence. The spread in the flight times of fragment ions results not only from this initial thermal motion of the molecules, but also from the release of kinetic energy in the decompositions that produce them. From the form of the time-of-flight coincidence spectrum, it is possible to deduce how much energy is released as kinetic energy in ionic dissociation[23], which is most important for the interpretation of dissociation mechanisms.

7.7.2 RESULTS

The first experiments involving the use of photoelectron–photoion coincidence spectroscopy[24] were measurements of photoelectron spectra in coincidence with particular ions, which can lead to the determination of the ion breakdown diagrams. One object of such

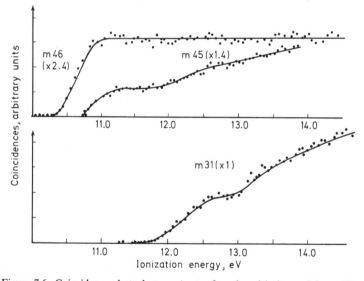

Figure 7.6. Coincidence photoelectron spectra for ethanol in integral form. These are integral photoelectron spectra of electrons detected in coincidences with the ions of the masses indicated. (From Brehm, Fuchs and Kebarle[25], by courtesy of Elsevier Publishing Company)

experiments is to test QET calculations, which also give breakdown diagrams as a primary result. An example is given in *Figure 7.6*, which shows the results of Brehm, Fuchs and Kebarle[25] on ethanol.

For experimental reasons, a retarding field photoelectron spectrometer was used, so the curves in *Figure 7.6* are in integral form. The complete differential photoelectron spectrum of ethanol would be obtained by adding the three curves together in the correct proportions and then differentiating the result. The parent ion, $C_2H_5OH^+$, is of mass 46, and the fact that the mass 46 coincidence curve remains flat above 11 eV means that all parent ions formed with higher energy dissociate completely. The fragment ion coincidence curve for mass 45 has a flat portion around 11.5 eV, which corresponds with a minimum in the photoelectron spectrum, that is, no parent ions are formed at this energy. The flat portion in the mass 31 curve at 13 eV corresponds with a second minimum in the photoelectron spectrum, which is not, however, very deep and some ions are still formed at this energy. The lack of a corresponding flat portion at 13 eV in the mass 45 curve is a contradiction of the statistical theory, because the formation of the two ions should be in competition. At 13 eV, fragment ions of mass 45 are being formed but no ions of mass 31, although there is sufficient energy available for both processes to take place.

When investigating the mechanism of an ionic dissociation, it is more useful to choose a single electron energy and to examine the identities and kinetic energies of the fragment ions formed. Experiments of this nature have been carried out only very recently, as they require a differential type of electron energy analyser whose poor electron collection efficiency can cause a serious intensity problem in coincidence measurements. In order to interpret the results of such experiments, the analysis of the photoelectron spectrum is a first requirement and a knowledge of the dissociation limits for the different fragmentation processes is also vital. On both counts it has seemed most profitable to first investigate very small molecules for which the necessary information is available.

Carbon tetrafluoride

The first band in the photoelectron spectrum of carbon tetrafluoride is continuous; the only energetically possible fragmentation products are $CF_3^+ + F$ in their ground states, and the correlation rules here allow the possibility of a direct dissociation (Section 7.4.1). The position of the dissociation limit is rather uncertain, but a good estimate is 14.2 eV[26]. The equivalent ionization energy at a point early in the first band in the photoelectron spectrum is 15.92 eV, so the excess energy available there is 1.72 eV. CF_3^+ ions detected in coincidence with photoelectrons from this point[27] are found to have a kinetic energy of 0.24 eV; the implication of this result for the total kinetic energy release must next be deduced.

From conservation of momentum and energy, it follows that in any decomposition into two fragments the fraction of the total kinetic energy release carried by each fragment is proportional to the mass of the other. In the dissociation of a molecule AB into A and B:

$$T_A = (m_B/m_{AB})T$$

or

$$T_B = (m_A/m_{AB})T \tag{7.28}$$

where T is the total kinetic energy release and m_A, m_B, etc., are masses. From the 0.24 eV of kinetic energy carried by CF_3^+ ions, the total kinetic energy release can be calculated by using equation 7.28 and is 1.10 eV. This value is 64% of the available excess energy and immediately suggests direct dissociation, but before drawing a definite conclusion one must consider how much translational energy is expected. In the fragmentation of an excited diatomic molecule, all of the excess energy appears as kinetic energy, as the fragments have no internal degrees of freedom, except electronic. This is not so in polyatomic dissociations and even in direct dissociation it is expected that a part of the available energy will go into vibration and rotation of the diatomic or polyatomic products. The minimum fraction expected to go into translation can be calculated by using a mechanical model of direct dissociation.

The molecule is considered as a triatomic molecule ABC, where B–C is the bond to be broken, and it is assumed further that the whole of the excess energy available, $E_{avl.}$, is released in breaking this bond and that atom B can move independently of A. The translational energy taken by C is $(m_B/m_{BC})E_{avl.}$, while the atom B has kinetic energy $E_{avl.}(m_C/m_{BC})$ with which it collides with A. The kinetic energy of AB then becomes $E_{avl.} m_B m_C/(m_{BC}m_{AB})$. The total energy going into translation of AB and C is therefore

$$T = \frac{m_B}{m_{BC}} \cdot \frac{m_{ABC}}{m_{AB}} \cdot E_{avl.} \tag{7.29}$$

This calculation assumes that the bond between A and B is so soft that B can move independently of A; if the A–B bond is absolutely rigid and the dissociation proceeds from and in a linear configuration, then the whole of the available energy must go into translation, as in the decomposition of a diatomic molecule. In real direct dissociations of polyatomic ions, the total kinetic energy release should be in the range between 100% of the available energy and the minimum given by equation 7.29.

When carbon tetrafluoride is treated as a triatomic F_3–C–F molecule, the minimum energy going into translation in a direct dissociation is calculated from equation 7.29 to be 50% of the available energy. The observation of 64% of the available energy as translational energy therefore supports the idea that this dissociation of CF_4^+ is direct, which is further confirmed by the fact that the distribution of kinetic energies is very sharp. When higher ionization energies within the first electronic state of CF_4^+ are chosen, however, it is found that the kinetic energy release remains effectively constant, and so the fraction of the available energy released as kinetic energy decreases. It seems that vibrational energy within the $^2T_1$ state of CF_4^+ contributes little to the kinetic energy of fragmentation, which is consistent with the fact that the t_1 orbital of CF_4 has no density whatever at the carbon atom and is fully C–F non-bonding. The vibrations that are initially excited on ionization cannot correspond to motion in the reaction co-ordinate, and as the fragmentation is very rapid they must go over to vibrations of the CF_3^+ product ion. Hence the dissociation of the ground state of the CF_4^+ ion is direct as far as the electronic part of the energy is concerned, but could also be considered as a vibrational pre-dissociation on a single potential energy surface.

Methyl chloride

The second band in the photoelectron spectrum of methyl chloride has already been mentioned as a possible example of a band broadened by direct dissociation (Section 7.4.1) and the coincidence measurements bear this out[27]. At the onset of the second band, the CH_3^+ ions produced have a kinetic energy of 0.28 eV with a narrow distribution, corresponding to 0.40 eV total energy release or 60–90% of the available energy; the large uncertainty in this percentage stems from doubt about the position of the true dissociation limit. For comparison, the direct model (equation 7.29) would allow between 85 and 100% of the energy available for this dissociation to appear in the form of translation. As higher excitation energies within the $^2A_1$ band are chosen, it is found that the release of fragment kinetic energy increases, but not as rapidly as the internal energy. It seems that the electronic excitation energy of $CH_3Cl^+(^2A_1)$ goes over directly to kinetic energy of the fragments and part of the vibrational excitation is also useful for breaking the bond. This is another instance where the normal modes that are excited on ionization do not correspond completely to motion in the reaction co-ordinate, although the agreement is closer than in the dissociation of CF_4^+ ($^2T_1$). This is expected, because the electron ionized in forming the $^2A_1$ state comes from an orbital that has mainly C–Cl bonding character.

Nitrous oxide

The photoelectron spectrum of nitrous oxide contains four bands[30], which correspond to the states $X\,^2\Pi$, $A\,^2\Sigma^+$, $B\,^2\Pi$ and $C\,^2\Sigma^+$ of N_2O^+. The lowest energy dissociation leads to the fragments $NO^+\,(^1\Sigma^+) + N\,(^4S_u)$, which, according to the correlation rules, give only a $^4\Sigma^-$ state, so that the formation of these products from any of the doublet states of N_2O^+ is spin forbidden. There are a number of other dissociation limits for different products in ground and electronically excited states, shown in *Table 7.2*, but there is no allowed fragmentation for the A state. Fluorescence emission is known to occur from the A state and the radiative lifetime has been measured directly[28].

Table 7.2 STATES OF N_2O^+ AND ITS DISSOCIATION PRODUCTS

Energy, eV	Substance(s) and state(s)	States obtained by combining the products
12.93	$N_2O^+\,(X\,^2\Pi)$	
14.17	$NO^+\,(^1\Sigma^+) + N\,(^4S_u)$	$^4\Sigma^-$
15.30	$O^+\,(^4S_u) + N_2\,(^1\Sigma_g^+)$	$^4\Sigma^-$
16.39	$N_2O\,(A\,^2\Sigma^+)$	
16.56	$NO^+\,(^1\Sigma^+) + N\,(^2D_u)$	$^2\Sigma^-,\ ^2\Pi,\ ^2\Delta$
17.25	$N_2^+\,(^2\Sigma_g^+) + O\,(^3P_g)$	$^{2,4}\Sigma^-,\ ^{2,4}\Pi$
17.65	$N_2O^+\,(B\,^2\Pi)^*$	
17.75	$NO^+\,(^1\Sigma^+) + N\,(^2P_u)$	$^2\Sigma^+,\ ^2\Pi$
18.59	$N_2^+\,(^2\Sigma_g^+) + O\,(^1D_g)$	$^2\Sigma^+,\ ^2\Pi,\ ^2\Delta$
19.02	$O^+\,(^2D_u) + N_2\,(^1\Sigma_g^+)$	$^2\Sigma^-,\ ^2\Pi,\ ^2\Delta$
19.47	$N^+\,(^3P_g) + NO\,(^2\Pi)$	$^{2,4}\Sigma^+,\ ^{2,4}\Sigma^-,\ ^{2,4}\Pi,\ ^{2,4}\Delta$
20.11	$N_2O\,(C\,^2\Sigma^+)$	

* The B state gives a broad band in the photoelectron spectrum reaching up to 19 eV.

Coincidence measurements[29] show that N_2O^+ ions formed in the $A\,^2\Sigma^+$ state with no vibrational excitation are indeed all detected as N_2O^+ ions, that is, they all fluoresce to the stable ionic ground state $X\,^2\Pi$. However, when N_2O^+ ions are formed in the A state with vibrational excitation energy (one quantum in v_1 or v_3), only some of them fluoresce (about 60%) while the remainder dissociate and appear as NO^+ ions with high kinetic energy. This is therefore one of the rare instances of competition between emission and pre-dissociation. From the radiative lifetime of the N_2O^+ ions in $A\,^2\Sigma^+(1,0,0)$ and the percentages of emission and pre-dissociation, the pre-dissociation lifetime can be calculated to be 0.4×10^{-6} s. The pre-dissociation is very slow, as befits a process that is forbidden by the spin rule and also by the $\Sigma^+ \leftrightarrow \Sigma^-$ rule. The total kinetic energy released in formation of the NO^+ ions is found[27] to be sharply distributed, which suggests that the mechanism of the pre-dissociation involves the crossing of potential energy surfaces at a single critical configuration.

For N_2O^+ ions in the third excited state, $C\,^2\Sigma^+$, *Table 5.2* shows that there are three energetically possible fragmentations that are fully allowed by the correlation rules, and six others that are forbidden only by the usually weak $\Delta\Lambda = 0$ selection rule. From such a plethora of possibilities one might be forgiven for expecting that at least one would be sufficiently favourable to lead to a direct dissociation. That this is not in fact so is shown, however, both by the photoelectron spectrum[30], in which the C band has well resolved fine structure, and by coincidence experiments[27], in which at least two competing fragmentations of C ions have been found. This is an indication of the complexity of the problem of explaining dissociations of even small molecular ions.

Carbon disulphide

As a final example of results from photoelectron–photoion coincidence spectroscopy, the dissociations of CS_2^+ ions in the D state

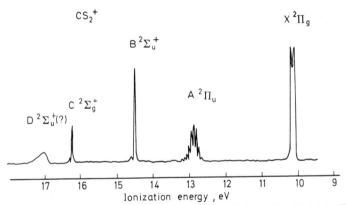

Figure 7.7. Photoelectron spectrum of carbon disulphide ionized by He I light. The two peaks in the first band at 10 eV are produced by spin–orbit coupling

is considered. The D state corresponds to the fifth band in the photoelectron spectrum (*Figure 7.7*), a band which has no counterpart in the spectra of carbon dioxide, nitrous oxide or carbonyl sulphide and which is continuous, in contrast to all other bands in the photoelectron spectra of linear triatomics hitherto observed.

A possible origin of the continuous band is ionization from the sulphur 3s-based σ_u orbital to give a $^2\Sigma_u^+$ state[30]; this explanation is supported by the large increase in intensity of D relative to the other bands on changing from He I to He II excitation[31]. An alternative origin might be a two-electron process leading to the electron configuration $\sigma_g^2\sigma_u^2\pi_u^4\pi_g^2\pi_u$, which gives ionic states $^4\Pi_u$, $^2\Phi_u$ and three $^2\Pi_u$ states if the molecule remains linear. The molecular ion might be bent in the equilibrium position after the two-electron process, however, and the consequent excitation of the low-frequency bending mode with Renner–Teller complications could explain the continuous appearance of the band. Another explanation of its lack of structure is a short lifetime of the CS_2^+ ions in the D state, and this possibility can be investigated by photoelectron–photoion coincidence spectroscopy. If the band is continuous because of rapid decomposition, the dissociation must be direct and should lead predominantly to one set of products with a large fraction of the available energy going into translation. The coincidence experiments show the opposite effect[27]. All of the three energetically possible product ions are formed by dissociations from $CS_2^+(D)$, namely S_2^+, CS^+ and S^+ with relative intensities of 0.1, 0.6 and 0.3, respectively. The kinetic energies of the S^+ and CS^+ ions are just large enough to prove that these fragments are formed in their electronic ground states, and the fractions of the available energy released into translation are only 35 and 22%. A direct dissociation mechanism is excluded as it would require between 50 and 100% of the excess energy to appear in translation and a single product to be formed predominantly. The fact that S_2^+ ions are produced at all is a clear indication that the dissociation is slow, as the transition state that leads to their formation must differ very considerably from the linear configuration reached in photoionization. Fast dissociation can be excluded as a reason for the lack of structure in the fifth photoelectron band of carbon disulphide, and the alternatives must now be examined. Apart from the two electron process proposed above, another, perhaps more likely, possibility is a short lifetime not towards decomposition, but towards internal conversion from D to C.

7.8 ION KINETIC ENERGY SPECTROSCOPY

Photoelectron spectrometers normally measure the kinetic energy spectra of electrons, but it is a simple matter to reverse the electrical potentials and to record kinetic energy spectra of the positive ions.

It is clear from the preceding examples that the kinetic energies released in fragmentation are an important indicator of dissociation mechanisms, and even without using coincidence techniques to select particular initial states, such kinetic energy spectra can be very useful. Because parent molecular ions have only random

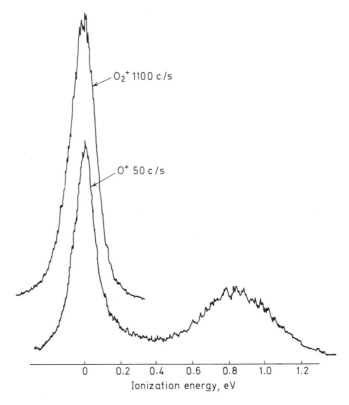

Figure 7.8. Energies of positive ions from ionization of oxygen with He I light. The upper curve is for O_2^+, the lower curve for O^+

thermal velocities, an electric field in the ionization region is needed in order to accelerate them into the energy analyser. The velocities of the ions in the analyser are proportional to the square root of their kinetic energy and inversely proportional to the square root of their masses, so if the ionizing light is pulsed* their masses can be determined by measurement of their time of flight. The kinetic

* A normal capillary discharge in helium can be pulsed satisfactorily by connecting the lamp as a relaxation oscillator and adjusting the helium pressure.

energy spectrum of the ions of each mass can therefore be measured separately.

As an example of such a measurement, the kinetic energy spectrum of O^+ ions from oxygen photoionized by light of 584 Å wavelength is shown in *Figure 7.8*. Oxygen atomic ions are found in two groups with energies of about 0.8 and 0 eV. Because the masses of O^+ and O are equal, the formation of O^+ ions with an energy of 0.8 eV corresponds to a release of a total of 1.6 eV, which is exactly equal to the energy available for the formation of the products in their ground states from O_2^+ ions in the $B\ ^2\Sigma_g^-$ state (see *Figure 1.6*). These ions therefore originate from pre-dissociation of the B state.

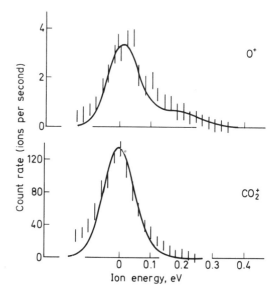

Figure 7.9. Kinetic energies of the CO_2^+ and O^+ ions from carbon dioxide ionized by He I radiation. (From Eland[32], by courtesy of Elsevier Publishing Company)

The O^+ ions with almost zero kinetic energy must originate either from a dissociation of $B\ ^2\Sigma_g^-$ oxygen ions to produce electronically excited products, or from pre-dissociation of O_2^+ ions formed initially in the $b\ ^4\Sigma_g^-$ state in which some vibrational levels just above the dissociation limit are populated by photoionization. Coincidence measurements have shown that the latter explanation is, in fact, correct[32].

Another example of the use of photoelectron kinetic energy analysis is given by the study of O^+ ions from carbon dioxide also ionized by He I radiation[33] (*Figure 7.9*). Although the ion energy resolution in *Figure 7.9* is not high, it is sufficient to prove that O^+ ions are formed with two distinct kinetic energies that correspond to total energy releases of about 0.06 and 0.32 eV. The energy level diagram in *Figure 7.10* shows that in the dissociation of vibrationless CO_2^+ from the state $C\,^2\Sigma_g^+$, O^+ is the only possible ionized product and the accompanying carbon monoxide molecule can be formed with either zero or one quantum of vibrational energy. The two kinetic

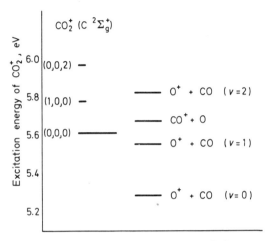

Figure 7.10. Energy levels of carbon dioxide ions in the $C\,^2\Sigma_g^+$ state compared with the dissociation limits for the formation of the possible products. (From Eland[32], by courtesy of Elsevier Publishing Company)

energy releases that are detected by ion kinetic energy spectroscopy correspond to these two vibrational states of carbon monoxide. The measurements lead, in conjunction with coincidence experiments, to a determination of the relative probabilities for the formation of CO ($v = 0$) and CO ($v = 1$) as about 15 and 85%, respectively. These figures are effectively Franck–Condon factors for predissociation, and their values are closely related to the forms of the potential energy surfaces involved.

REFERENCES

1. HERZBERG, G., *Spectra of Diatomic Molecules*, 2nd edn, Van Nostrand, New York, 413 (1950)
2. REID, N. W., *Int. J. Mass Spectrom. Ion Phys.*, **6**, 1 (1971)
3. LINDHOLM, E., in Franklin, J. (Editor) *Ion Molecule Reactions*, Plenum Press, New York (1972)

4. SETSER, D. W., *MTP International Review of Science, Physical Chemistry, Series One*, Volume 9, *Chemical Kinetics*, Butterworths, London (1972)
5. MULLIKEN, R. S., *J. chem. Phys.*, **33**, 247 (1960)
6. HERZBERG, G., *Electronic Spectra and Electronic Structure of Polyatomic Molecules*, Van Nostrand, Princeton, N.J. (1966)
7. HASTED, J. B., *Physics of Atomic Collisions*, Butterworths, London (1964)
8. CALVERT, J. G. and PITTS, J. N., *Photochemistry*, John Wiley, New York, 186 (1966)
9. KRONIG, R. DE L., *Z. Physik*, **50**, 347 (1928)
10. CZARNY, J., FELENBOK, P. and LEFEBVRE-BRION, H., *J. Physics*, **B4**, 124 (1971)
11. FROSCH, R. P., in Pitts, J. N. (Editor) *Excited State Chemistry*, Gordon and Breach, New York (1970)
12. ROSENSTOCK, H. M., in Kendrick, E. (Editor) *Advances in Mass Spectrometry*, Vol. 4, Institute of Petroleum, London, 523 (1968)
13. SAMSON, J. A. R., GARDNER, J. L. and MENTALL, J. E., *J. Geophys. Res.*, **77**, 5560 (1972)
14. FIELD, F. H. and FRANKLIN, J. L., *Electron Impact Phenomena and the Properties of Gaseous Ions*, Academic Press, New York (1970)
15. WAGMAN, D. D., EVANS, W. M., MALOW, I., PARKER, V. B., BAILEY, S. M. and SCHUMM, R. H., *N.B.S. Tech. Note No. 270–3*, U.S. Government Printing Office, Washington, D.C. (1968)
16. BRUNDLE, C. R., *Chem. Phys. Lett.*, **7**, 317 (1970)
17. RAFTERY, J. and RICHARDS, W. G., *J. Physics*, **B5**, 425 (1972)
18. SCHNEIDER, B. S. and SMITH, A. L., in Shirley, D. A. (Editor) *Electron Spectroscopy*, North Holland, Amsterdam, 335 (1972)
19. HAUGH, M. J. and BAYES, K. D., *J. phys. Chem.*, **75**, 1472 (1971)
20. BAKER, C. and TURNER, D. W., *Proc. R. Soc., Lond.*, **A308**, 19 (1968)
21. BERKOWITZ, J., CHUPKA, W. A. and WALTER, T. A., *J. chem. Phys.*, **50**, 1497 (1969)
22. SCHAEFER, H. F., III and HEIL, T. G., *J. chem. Phys.*, **54**, 2573 (1971)
23. ELAND, J. H. D., *Int. J. Mass Spectrom. Ion Phys.*, **8**, 143 (1972)
24. BREHM, B. and VON PUTTKAMER, E., *Z. Naturforsch.*, **22a**, 8 (1967)
25. BREHM, B., FUCHS, V. and KEBARLE, P., *Int. J. Mass Spectrom. Ion Phys.*, **6**, 279 (1971)
26. NOUTARY, C. J., *J. Res. natn. Bur. Stand.*, **72A**, 479 (1968)
27. BREHM, B., ELAND, J. H. D., FREY, R. and KUSTLER, A., *Int. J. Mass Spectrom. Ion Phys.*, to be published
28. FINK, E. H. and WELGE, K. H., *Z. Naturforsch.*, **23a**, 358 (1968)
29. ELAND, J. H. D., *Int. J. Mass Spectrom. Ion Phys.*, to be published
30. BRUNDLE, C. R. and TURNER, D. W., *Int. J. Mass Spectrom. Ion Phys.*, **2**, 195 (1969)
31. PRICE, W. C., POTTS, A. W. and STREETS, D. G., in Shirley, D. A. (Editor) *Electron Spectroscopy*, North Holland, Amsterdam, 187 (1972)
32. ELAND, J. H. D., *Int. J. Mass Spectrom. Ion Phys.*, **9**, 397 (1972)

8 Applications in Chemistry

8.1 INTRODUCTION

It may be helpful to categorize applications of photoelectron spectroscopy into two major groups, those in which the technique is used as a tool, as in analysis and the study of unimolecular reactions, and those in which the spectra themselves are studied for the insight they give into molecular electronic structure. The uses of photoelectron spectroscopy in the study of reaction mechanisms and mass spectrometry have been considered at length in Chapter 7, and of the first group of applications those in analysis remain to be discussed here. The greater part of this chapter is devoted to chemical applications and to problems of molecular electronic structure and chemical bonding. Selection has inevitably had to be exercised in choosing the material, and the author has attempted to consider problems that might be said to have existed in chemistry before photoelectron spectroscopy was invented to solve them. The reference list has been made as complete as possible so that important points can be followed up and the spectra of the many compounds referred to can be found.

8.2 d ORBITALS IN CHEMICAL BONDING

Photoelectron spectroscopy has not yet penetrated very far into the homeland of d orbitals, the complexes of the transition metals, mainly because the majority of such compounds are very involatile. This is unfortunate, because molecular orbital diagrams, which photoelectron spectroscopy could provide experimentally, are important for an understanding of their chemistry. Molecular orbital theory and ligand field theory provide qualitative predictions of the form of the orbital diagrams, and other spectroscopic methods give experimental information about the metal d orbitals in particular. The photoelectron spectra of complexes, interpreted by using Koopmans' theorem, could provide complete molecular orbital diagrams that would show not only the metal d orbitals but also the ligand orbitals and how the d, σ and π orbitals stand in relation to one another. Apart from the transition metal complexes, the involvement of d orbitals in the bonding of non-metal compounds, $d\pi$–$p\pi$ bonding, has been a controversial question for a long time, and here photoelectron spectroscopy is able to make an important contribution.

8.2.1   TRANSITION METAL COMPLEXES

The most volatile transition metal complexes are the mainly covalent complexes with σ-donating and π-accepting ligands such as CO or PF_3 or with organic π systems, and several such compounds have been studied. They include the carbonyls[1, 2], a series of pentacarbonylmanganese derivatives[3], several π-arene complexes[4], some PF_3 derivatives[5, 6], enolate anion adducts[7, 8] and bis-(π-allyl) complexes[9]. As examples of the spectra of this type of compound, those of some carbonyls and trifluorophosphine adducts will be discussed here. The bonding in these complexes involves the donation of σ electrons from the ligands into empty metal orbitals, accompanied by back-donation from metal d orbitals into empty ligand π^* orbitals of appropriate symmetry. These effects are co-operative, as the more charge is transferred in one direction by one mechanism the greater is the inducement for it to be replaced by the other mechanism, thus restoring electroneutrality. The ligands must be both σ donors and π acceptors, and of these properties that of accepting π electrons seems to be the most important for the stability of the compounds. The molecular field produced by the ligands is strong, and the complexes are almost all of low spin. The metals that form these complexes occur in the second half of the transition series and are much more common in the first than in the second and third transition series. These are the metals that have several occupied d orbitals of relatively high energy from which donation can take place.

For the interpretation of the photoelectron spectra, these qualitative ideas must be translated into observable effects on molecular orbital energies. The orbitals involved are the filled d and empty s and p atomic orbitals for the metal, and the full σ and empty π^* orbitals for the ligands. Compound formation should cause stabilization of both the ligand σ orbitals and the metal d orbitals compared with their energies in the metal atom and free ligand molecules. If problems of local charges and Koopmans' approximation are neglected, this should lead to an increase in both sets of ionization potentials, visible in the photoelectron spectra.

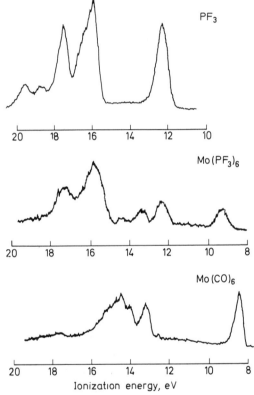

Figure 8.1. He I photoelectron spectra of phosphorus trifluoride, molybdenum hexakistrifluorophosphine and molybdenum hexacarbonyl

Figure 8.1 shows the photoelectron spectra of $Mo(CO)_6$, $Mo(PF_3)_6$ and of the ligand PF_3. The band at lowest ionization potential in the spectra of the two complexes shifts by only 0.7 eV when the ligand is changed and must be attributed to ionization from the metal d orbitals. As the complexes are of low spin, all of the six d electrons are in t_{2g} orbitals and the d electron ionization bands are single. The electron configuration of atomic molybdenum and that of the molybdenum in the complexes are so different that the d electron ionization potentials in the two states cannot be compared directly. The metal d orbital ionization potentials in the CO and PF_3 complexes can be compared, however, and as PF_3 causes the greater stabilization, it seems to be a better π acceptor than CO.

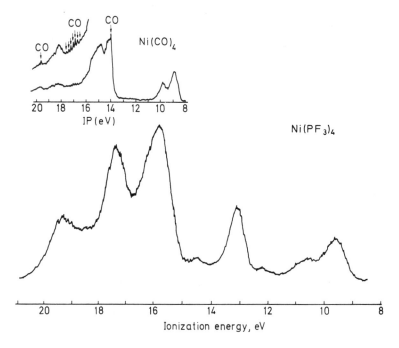

Figure 8.2. He I photoelectron spectrum of nickel carbonyl and nickel tetrakistri-fluorophosphine. (Spectrum of nickel carbonyl by courtesy of D. R. Lloyd)

In the high ionization potential region between 15 and 20 eV there is little difference between the photoelectron spectrum of PF_3 and that of its molybdenum adduct. This means that compound formation has only a weak effect on the energy of the fluorine lone-pair orbitals or on the P–F bonding orbitals. The band at 12.2 eV in the PF_3 spectrum represents ionization from the lone pair located mainly on the phosphorus atom, the orbital from which σ donation occurs. In octahedral symmetry, the six ligand lone-pair orbitals become t_{1u}, e_g and a_{1g} molecular orbitals, which can be stabilized by interaction with empty molybdenum p, d or s orbitals, respectively. The three bands in the $Mo(PF_3)_6$ spectrum at 12.4, 13.5 and 14.5 eV are identified by their relative areas with ionization from these three orbitals in the order given. It can be seen that all three orbitals are stabilized relative to the lone pairs of free PF_3, in accordance with their bonding character. In the $Mo(CO)_6$ spectrum, only one band in the σ bonding region can be seen (13.2 eV) and this band probably represents the t_{1u} orbital. The other σ bonding orbitals presumably overlap with the part of the spectrum attributed to ionization from orbitals located in the CO molecules.

Figure 8.2 shows the photoelectron spectra of $Ni(PF_3)_4$ and $Ni(CO)_4$. As nickel in the zero oxidation state has a completed d shell, both the t_2 and e orbitals of these tetrahedral complexes are full. The first bands again represent d electron ionizations and are split into two components with approximate relative areas of 3:2, as expected for t_2^{-1} and e^{-1} ionizations. The splitting between the states $^2T_2$ and ^2E, which crudely indicates the strength of the ligand field, is about the same in the PF_3 and CO complexes, but once again the ionization potential of the PF_3 compound is higher, showing that PF_3 is a better π acceptor than CO. The analysis of the remainder of the $Ni(PF_3)_4$ spectrum is similar to that given for $Mo(PF_3)_6$; again, the PF_3 lone-pair σ-donating orbitals are stabilized in the complex.

The photoelectron spectra of these complexes illustrate a valuable generalization that was made by Evans *et al.*[3] in their work on the pentacarbonylmanganese derivatives, namely that the spectra of the metal complexes can often be divided into three regions. In order of increasing ionization potential come the metal d orbitals, the metal-to-ligand bonding orbitals and then pure ligand orbitals. This rule is useful in interpreting the photoelectron spectra and accords well with the fact that the metal d orbitals determine many of the chemical properties of the complexes.

8.2.2 $d\pi$–$p\pi$ BONDING

Many 'anomalous' properties of compounds of the elements from silicon to sulphur with atoms that contain electrons in orbitals of π symmetry can be explained by the assumption of $d\pi$–$p\pi$ bonding, and if any chemical or physical property varies discontinuously between the first and second rows on descending Group IV, V or VI of the Periodic Table, $d\pi$–$p\pi$ bonding may be invoked as the cause of the variation. Properties that are often scrutinized for discontinuous variation are molecular geometry, bond energies or dipole moments, and although $d\pi-p\pi$ bonding is usually a reasonable explanation of the observations it is difficult to establish it unambiguously as the unique explanation. The presence of $d\pi-p\pi$ bonding should affect molecular orbital energies and orbital bonding properties and so be observable in photoelectron spectra. Because the characters of individual orbitals are reflected in photoelectron spectra, it is possible to single out those orbitals whose symmetry properties allow them to involve d orbitals, and this procedure results in a more specific test of d orbital participation. The involvement of d orbitals can be established from photoelectron spectra in two ways. Firstly, vibrational structure may be found that indicates bonding character which the orbital concerned cannot possess, for symmetry reasons, without the participation of d orbitals. Alternatively, stabilization of a particular orbital that can interact with d orbitals may be found as an increase in the ionization potential compared with those from other orbitals whose symmetry makes it impossible for them to have d character. Unfortunately, resolved vibrational structure is not usually found in the spectra of molecules for which $d\pi-p\pi$ bonding is possible, and most of the results obtained so far involve the use of the second method. An exception is sulphur dioxide, whose photoelectron spectrum has been studied at high resolution[10]. The second band in the photoelectron spectrum contains overlapping progressions that belong to two electronic states of the ion, $^2B_2$ and $^2A_2$. According to the vibrational analysis, one of the orbitals involved, $5b_2$ or $1a_2$, is strongly bonding between S and O and the other is S–O non-bonding, while occupancy of both the orbitals has a strong effect on the bond angle. Now, according to the orbital symmetries, the $1a_2$ orbital is completely S–O non-bonding in the absence of sulphur d orbital participation, and $5b_2$ is only weakly S–O bonding. It appears that $d\pi-p\pi$ bonding must be effective in at least one of these orbitals in order to account for the S–O bonding character found. This is confirmed by an SCF molecular orbital calculation[11], which shows that the S–O bonding orbital is indeed $1a_2$, which has the lower ionization potential, and that it owes its bonding character to overlap with sulphur d orbitals.

A series of compounds in which the special symmetry properties of d orbitals play an important role are the cyclic phosphonitrilic halides, $(PNX_2)_n$:

The electronic structure of these compounds resembles that of the aromatic hydrocarbons in that it can be described in terms of separate

σ and π systems. All phosphonitrilic derivatives with $n = 3$ and the phosphonitrilic fluorides with $n = 3-6$ have their PN rings planar or nearly so, and all P–N bond lengths equal. Two π systems are distinguished, one in the plane of the ring (π_s) and the other out of the ring plane, as in aromatic hydrocarbons (π_a). The in-plane system is *homomorphic*, which means that in an orbital in which all overlaps are favourable, the p wave-functions on two consecutive nitrogen atoms have the same sign. The out-of-plane π system is *heteromorphic* because the symmetry of the d_{xz} orbitals involved means that consecutive nitrogen p wave-functions must have opposite signs for favourable overlap. The molecular energy diagram for a heteromorphic system is different from that for a homomorphic system, and a heteromorphic system can be aromatic with any number of delocalized electrons[12]. The phosphonitrilic halides are indeed aromatic for all values of n, as shown by their chemical properties. Branton et al.[13] have measured the photoelectron spectra of a series of phosphonitrilic fluorides and have also measured the first ionization potentials of other phosphonitrilic derivatives by an electron impact method; they compared their experimental ionization potentials with orbital energies for the two π systems calculated by a Hückel procedure. The ionization potentials agree with the calculated orbital energies sufficiently well to confirm the $d\pi$–$p\pi$ model, and the alternation in first ionization potentials with n shows that the homomorphic system is energetically outermost, in agreement with previous findings. Because of overlap between ionization bands from the two systems of π orbitals, less sharp structure is seen in the spectra of the phosphonitrilic fluorides than in those of aromatic hydrocarbons.

In a detailed investigation of $d\pi$–$p\pi$ bonding, Frost et al.[14] measured the photoelectron spectra of the halosilanes for comparison with those of the halomethanes. The spectra of chloroform and its silicon analogue trichlorosilane, are shown in *Figure 8.3* as an example. The spectra are very similar, and for every pair of isomorphic silicon and carbon compounds the order of the ionic states is the same; it is necessary to consider how they ought to differ in order to decide whether d orbital participation in bonding of the silicon compounds is important. Any d orbital effects should be most noticeable on the energies of ionization from the non-bonding halogen orbitals, as these are non-degenerate only because of non-bonding halogen–halogen interactions and weak interaction with the central atom orbitals[15].

In the absence of d orbital participation, the following changes are to be expected on going from carbon to silicon:

(1) the ionization potentials for all equivalent orbitals should be

reduced because silicon has a lower electronegativity than carbon;

(2) the lone-pair ionization bands should close up on one another because the greater size of the central atom should reduce the interactions between the lone-pair atomic orbitals;

(3) the lone-pair bands should become sharper for the same reason, as smaller changes in molecular geometry would follow ionization.

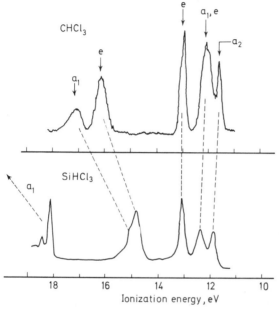

Figure 8.3. He I photoelectron spectra of chloroform and its silicon analogue. (SiHCl₃ from Frost *et al.*[14], by courtesy of the National Research Council of Canada)

In fact, the ionization potentials from equivalent lone-pair orbitals are all higher in chlorosilanes than in chloromethanes, although the ionization bands that involve σ bonding orbitals move to lower ionization potentials, as expected. There is no noticeable closing up or sharpening of the lone-pair bands, and in these respects the spectra of the carbon and silicon compounds are essentially the same. These deviations from the behaviour expected without d orbital participation are explained qualitatively if $d\pi$–$p\pi$ bonding is introduced. The $d\pi$–$p\pi$ interaction between chlorine and silicon can stabilize certain orbitals and increase their bonding character, so broadening the bands, but because of their different symmetries not all of the non-bonding orbitals enjoy these benefits equally. The differential stabilization maintains the separation between the bands, despite the reduced Cl–Cl non-bonding interactions following from the greater size of the silicon atom. A quantitative application of these ideas to the interpretation of the spectra is much more difficult. Frost et al.[14] used an empirical one-electron model, previously developed by Dixon et al.[15] for the halomethanes, but once d orbitals were introduced the number of parameters exceeded the number of measured ionization potentials. Nevertheless, by making some credible assumptions it was possible to obtain a good fit to the observed lone-pair ionization potentials with a set of physically reasonable parameters. The $d\pi$–$p\pi$ interaction per bond was found to decrease as the number of chlorine atoms increased, reflecting the decreasing ratio of silicon d orbitals to chlorine lone-pair orbitals. The effects observed are weaker for the fluorosilanes than for chlorosilanes, so that one might further deduce that Si–F bonding has less $d\pi$–$p\pi$ character than Si–Cl bonding.

Cradock and Whiteford[16] have measured the photoelectron spectra of all the monohalo- and dihalo-silanes and -germanes, and concluded from arguments similar to those given above that $d\pi$–$p\pi$ bonding is significant for them. The interpretation of their spectra led to the conclusion that $d\pi$–$p\pi$ bonding is not much less effective in the halogermanes than in the halosilanes and does not decrease by more than a factor of 2 or 3 on going from a chloride to the corresponding iodide. Green et al.[17] have measured the photoelectron spectra of many of the tetrahalides of Group IV elements, including also tin(IV) chloride and bromide. The same arguments applied to their spectra imply that $d\pi$–$p\pi$ bonding is also important for tin compounds, although perhaps less so than for those of silicon and germanium.

Other compounds whose photoelectron spectra have been interpreted as indicating $d\pi$–$p\pi$ bonding in the molecules include thiophene and its derivatives[18], trifluorophosphine oxide[19] and the halophosphines[20].

8.3 NON-BONDED INTERACTIONS

There are a number of electronic interactions between atoms or groups within molecules which are not represented by bonds in chemical formulae and are often omitted from simple molecular orbital models. They may nevertheless be strong and have an important effect on the electronic structure, particularly on the nature of the outermost orbitals. An example encountered in the preceding section was the mutual interaction between the halogen lone pairs in the halides of carbon and silicon. Other examples are the interactions of heteroatom lone pairs in large organic molecules, hyperconjugation and homoconjugation. These topics are all related to one another and have been studied extensively by photoelectron spectroscopy. For a discussion of the theoretical ideas involved, a review by Hoffmann[21] can be recommended.

8.3.1 HETEROATOM LONE PAIRS

In the photoelectron spectra of organic molecules that contain heteroatoms with lone pairs, the lone-pair ionization bands can often be identified either by their low ionization potentials or their relatively narrow contours. Very narrow bands are rarely seen, however, and if the molecules are conjugated, thus possessing occupied π orbitals that also have low ionization potentials, the lone-pair ionizations can be recognized only after model calculations and comparison with the photoelectron spectra of structurally related compounds. The surprising result that emerges is that if a molecule contains two equivalent lone pairs, there is not just one band in the photoelectron spectrum showing degeneracy of the lone-pair orbitals but two or more, even if the heteroatoms are separated by a number of bonds. Furthermore, the splitting does not decrease monotonically as the heteroatoms are placed farther away from one another in a series of compounds, but passes through a minimum and rises again.

An example of such splittings is provided by the photoelectron spectra of the sulphur ring compounds shown in *Figure 8.4*. There are considered to be two mechanisms that underlie this splitting, namely, *through-space* and *through-bond* interactions. If there are two equivalent lone pairs in a molecule described by atomic orbital wave-functions ϕ_1 and ϕ_2, then two symmetry-adapted molecular orbitals can be formed from the atomic orbitals:

$$\psi_s = 1/\sqrt{2}\,(\phi_1 + \phi_2) \tag{8.1}$$

$$\psi_a = 1/\sqrt{2}\,(\phi_1 - \phi_2) \tag{8.2}$$

If the atomic orbitals interact directly through space, the symmetrical combination ψ_s will have the lower energy (higher ionization potential) when the overlap integral $S_{ab} = <\phi_1 | \phi_2>$ is positive, but the unsymmetrical combination will have the lower energy if S_{ab} is negative. This through-space interaction will usually be

204

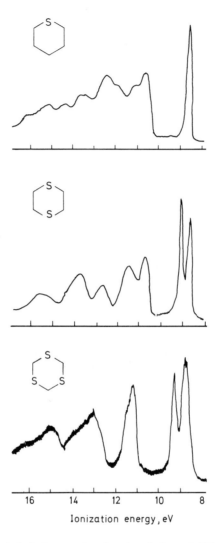

Figure 8.4. He I photoelectron spectra of pentamethylene sulphide, 1,4-dithiane and 1,3,5-trithiane. (From Sweigart, D. A. and Daintith, J., *Sci. Prog. Oxf.*, **59**, 325 (1971), by courtesy of Blackwells Scientific Publications Ltd.)

dominant if the two heteroatoms are neighbours, for instance, in the disulphides (S_{ab} positive), *trans*-azo compounds (S_{ab} negative), *cis*-azo compounds (S_{ab} positive), and it often remains dominant if they are separated by only one carbon centre. If the interaction is predominantly through-space, it should produce new orbitals with energies symmetrical about the energy of the original lone pairs or, more generally, the centre of gravity of the lone-pair ionization bands should not be shifted by this interaction alone.

The existence of through-bond interaction between lone pairs was first predicted by Hoffmann, Imamura and Hehre[22] by using extended Hückel theory. In particular, they found that for the diazabicycloalkanes,

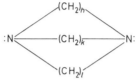

the through-space interaction with S_{ab} positive dominates when h, k and l are unity, but for h, k and l larger than unity a second-order interaction through the C–C and C–H bonding σ orbitals predominates. It is mainly an interaction with C–C σ orbitals of suitable symmetry and energy, and it effectively delocalizes the lone pair away from the nitrogen atoms over the rest of the molecule. This through-bond interaction makes the new orbital containing mainly a symmetrical combination of lone pairs energetically outermost, giving the opposite ordering to that expected on a through-space mechanism, and it also shifts the centre of gravity of the bands. The effects of the through-space and through-bond interactions on the lone-pair orbital energies are shown schematically in *Figure 8.5*.

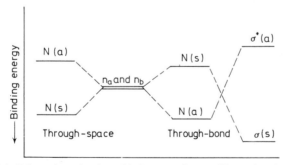

Figure 8.5. Schematic molecular orbital energy diagram illustrating the effects of through-space and through-bond interactions on the ionization potentials of lone-pair orbitals

The predictions of through-bond interaction for the case $h = k = l = 2$ (1,4-diazabicyclo[2.2.2]octane, DABCO) have been elegantly confirmed by Heilbronner and co-workers[23, 24]. Their photoelectron spectra of DABCO and two related compounds are shown in *Figure 8.6*, in which the splitting and shift to higher ionization potentials of the lone-pair bands in DABCO can be clearly seen.

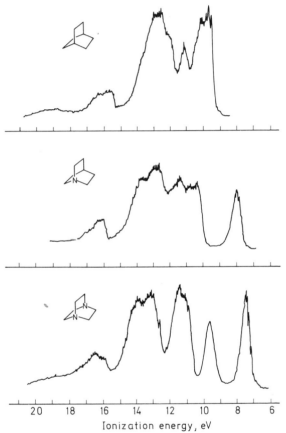

Ionization energy, eV

Figure 8.6. The photoelectron spectra of bicyclo[2.2.2]octane, quinuclidine and DABCO. (From Bischof *et al.*[23], by courtesy of Pergamon Press)

They were able to show, by an analysis of the vibrational fine structure of the two bands in terms of the structural changes following ionization, that the outermost orbital is indeed the symmetrical combination. The splitting between the two lone-pair ionization bands in DABCO and the shift of the centre of gravity of the two bands relative to the ionization energy of the lone pair in quinuclidine had been calculated theoretically as 1.6 and 0.37 eV, respectively, and these predictions were confirmed by experiment, where values of 2.1 and 0.57 eV, respectively, were found. The same effects can be seen qualitatively in the spectrum of 1,4-dithiane compared with that of pentamethylene sulphide (*Figure 8.4*).

The lone-pair interactions in nitrogen compounds have been studied much more than interactions that involve other heteroatoms. Compounds that have been investigated include several with azo groups, both *cis-* and *trans-*substituted[25–27], azabenzenes[28, 29], ethylenediamine[23], piperazine[23] and some hydrazines[27]. The interpretation of the spectra generally follows from comparison with empirical calculations that have been suitably parameterized. The analysis of the photoelectron spectra of the azabenzenes is still to some extent controversial, but the proper inclusion of the lone-pair interactions is clearly of crucial importance. The splittings between the lone-pair bands in the spectra of the diazines according to Gleiter *et al.*[28] are 2.0 eV for the *ortho* compound (pyridazine), 1.5 eV for the *meta* compound (pyrimidine) and − 1.7 eV for the *para* compound (pyrazine). These values show an alternation in magnitude which is characteristic of the transition from through-space to through-bond interaction, and the negative sign shows the consequent reversal in the order of the symmetrical and antisymmetrical orbital combinations.

The interactions of equivalent oxygen lone pairs in dicarbonyl compounds have been examined, and it has been shown that the through-bond interaction is dominant[30]. Calculated lone-pair orbital energy splittings agree well with the measured photoelectron spectra in instances when there is no doubt about the band identifications. The oxygen and sulphur lone-pair ionizations in some cyclic ethers and sulphides have also been studied by photoelectron spectroscopy[31] and the observed splittings can be interpreted in a similar manner. The through-space interaction between sulphur atoms seems to be stronger than that between oxygen atoms, perhaps because of the greater size of the sulphur orbitals, whereas the reverse may be true of the through-bond interaction. The magnitude of both effects must depend markedly on molecular geometry, and in an open-chain molecule the different accessible conformations may have very different interactions and so give rise to overlapping structure in the lone-pair region of the photoelectron spectrum. For this reason, it is more profitable to study lone-pair interactions in the relatively rigid ring compounds. Numerical values of the splittings between the lone-pair ionizations of a few interesting molecules are given in *Figure 8.7*.

The lone-pair interactions between equivalent halogen atoms, which each have two long pairs, is slightly more complicated than that discussed above. As mentioned in the section on spin–orbit coupling (Chapter 6, Section 6.2), a single halogen atom will usually give two lone-pair peaks in the photoelectron spectrum, either because of spin–orbit coupling or because one halogen p orbital interacts more strongly than the other with the remainder of the molecule. In a compound that contains two halogen atoms, four lone-pair ionization bands should be seen, although some components may

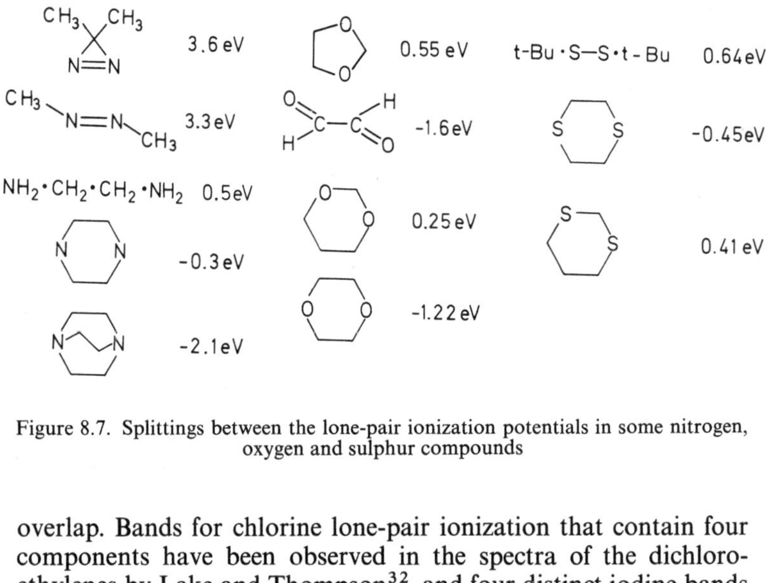

Figure 8.7. Splittings between the lone-pair ionization potentials in some nitrogen, oxygen and sulphur compounds

overlap. Bands for chlorine lone-pair ionization that contain four components have been observed in the spectra of the dichloro-ethylenes by Lake and Thompson[32], and four distinct iodine bands are found in the spectrum of methylene iodide. The dichloro-propanes and 1,4-dichlorobutane have been investigated[33]; their spectra contain broadened chlorine lone-pair bands, but in these instances the individual components are not resolved. The through-space and through-bond mechanisms must be expected to apply to halogen lone pairs just as to those of nitrogen, oxygen or sulphur, even though the appearance of the bands in the spectra is much more complicated.

Lone-pair interactions in compounds of phosphorus and the other heavier non-metals have not yet been investigated. Interaction between non-equivalent lone pairs, including lone pairs on different atoms, also remain a subject for future research.

8.3.2   ISOLATED DOUBLE BONDS

Apparently isolated π orbitals can interact by the same mechanisms as lone pairs and for them the through-space interaction is often referred to as *homoconjugation*, while *hyperconjugation* is one form of through-bond interaction. Photoelectron spectra of many of the most interesting compounds in this connection were measured first at very low resolution[34], but the details of the non-bonded interactions have been deduced from more recent high resolution work[35].

Model compounds for the through-space interactions of π bonds are the cyclic molecules norbornadiene, barrelene and *cis,cis,cis*-1,4,7-cyclononatriene, of which the last is also the archetypal *homoaromatic* compound. In the spectra of all these molecules, large splittings between the π ionization bands have been found[35–37]

Norbornadiene   Barrelene   *cis, cis, cis*-
                                            1,4,7-Cyclononatriene

with the ordering predicted for predominantly through-space interactions. Typical overlap integrals for the non-bonded interactions in these and similar molecules in an HMO treatment would be about 1 eV, compared with 2–3 eV for normal double bonds. In barrelene and *cis,cis,cis*-1,4,7-cyclononatriene there are three equivalent π bonds, and when these are combined one of the orbitals obtained is doubly degenerate (e) and the other is single (a), as shown in the qualitative molecular orbital diagrams below.

When electrons are removed from the e orbitals, ^2E ionic states result, which are liable to Jahn–Teller distortions and to splittings

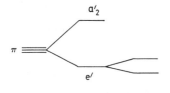

Barrelene

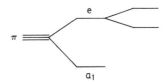

cis, cis, cis-
1,4,7-Cyclononatriene

in the spectra. The observation of such splitting in the second band of the photoelectron spectrum of barrelene and in the first band of the cyclononatriene, together with the greater intensities of the same bands, amply confirms the theoretical orderings. Another effect is observed in the spectra of barrelene and related molecules: as the number of double bonds is increased, the onset of the σ electron ionization bands moves to higher ionization potentials approximately linearly with the number of π bonds. This effect is rationalized by an increase in the 2s character of the σ bonds as more atomic p orbitals are taken over for π bonding. The replacement of hydrogen atoms by fluorine and the introduction of nitrogen atoms into aromatic rings have similar effects, shifting the σ levels more than the π levels to higher ionization potentials. The centre of gravity of the ionization system does move to a higher ionization potential as the number of double bonds, fluorine atoms or nitrogen atoms in a molecule is increased, but much less than does the onset of the σ ionization system. The π ionization band-shifts in barrelene and related molecules have been interpreted as evidence that through-bond interactions are also not negligible[36].

The observation of a splitting of about 0.95 eV between the π orbitals in *cis,cis,cis*-1,4,7-cyclononatriene is a most interesting result, as previous studies of the bond lengths, bond angles, nuclear magnetic resonance spectrum and heat of hydrogenation gave no evidence of homoaromaticity. The molecule has six π electrons and is identical with benzene except for the reduced overlap between alternate double bonds. Bischof, Gleiter and Heilbronner[37] showed theoretically that if the resonance integral between separate double bonds is given a value $m\beta$, where β is the resonance integral within the normal π bonds, the usual tests for aromaticity or delocalization will detect almost nothing unless m has a value greater than 0.3. The splitting of the photoelectron ionization bands, on the other hand, is directly proportional to m, and from the observed splitting of the π bands in the spectrum an m value of 0.26 was deduced. Photoelectron spectroscopy is therefore a much more sensitive test for the existence of orbital interaction than other physical measurements. On the other hand, the results must be treated with caution as they refer strictly to the ions and not the molecules. The conjugation may be much stronger in the ionized state than in the neutral species.

Because both the π orbitals in norbornadiene are singly degenerate, the orbital order could not be inferred directly from the photoelectron spectrum to confirm the predominantly through-space nature of the interaction. The proof of the orbital ordering is a very elegant example of a general method which Heilbronner[38] has proposed for the determination of such orderings. Isopropylidenenorbornadiene,

has almost the same dihedral angle between the two equivalent π bonds as norbornadiene itself and it has a new π bond, which, because of its symmetry, can interact only with the negative overlap combination of the two equivalent localized π orbitals. In the photoelectron spectrum of isopropylidenenorbornadiene[39], the higher ionization potential π band of the parent compound recurs at exactly its original energy. The first band in the spectrum of norbornadiene, on the other hand, is split strongly into two separate bands in the spectrum of its isopropylidene derivative. This effect is proof that the outermost π orbital in norbornadiene is the negative overlap combination, as required by the through-space interaction model.

The simplest model compound for the study of through-bond interactions of isolated π bonds is 1,4-cyclohexadiene, which has a large dihedral angle of about 160 degrees between the double bonds, and is of ideal symmetry for through-bond, hyperconjugative interaction. That the observed splitting of the π ionization bands in the spectrum of 1,4-cyclohexadiene is 1.0 eV[35], that is, greater than in norbornadiene (0.85 eV), is immediately surprising on a through-space interaction mechanism, as the π bonds are further apart. A proof that the symmetrical orbital combination has a lower ionization potential than the unsymmetrical combination, as consonant with a through-bond interaction, has been obtained by a similar method to that used for norbornadiene[38]. This result is also confirmed by a study of the spectrum of 1,4,5,8-tetrahydronaphthalene,

which contains two 1,4-cyclohexadiene rings.

The parameters needed to represent the ionization potentials of 1,4-cyclohexadiene correctly in an empirical calculation on the through-bond interaction model also led to a satisfactory representation of the photoelectron spectrum of the tetrahydronaphtha-lene[40]. The photoelectron spectra of many other molecules that contain apparently isolated double bonds have been examined and found to show splittings of the π orbital ionization bands, which are attributed to the two types of non-bonded interaction[41]. These interactions are not, however, restricted to lone-pair orbitals and π orbitals, but occur generally. This is an area in which experimental studies together with empirical molecular orbital calculations are greatly enriching the knowledge of molecular electronic structure.

8.4 SUBSTITUENT EFFECTS ON IONIZATION POTENTIALS

Photoelectron spectroscopy, like almost all other spectroscopic techniques before it, provides a method of measuring substituent effects in organic molecules. It is doubtful whether organic chemists really need a new method of measuring these effects, but the application of linear free energy relationships to ionization potentials is still useful. It can bring together a great deal of data into simple equations which allow unknown ionization potentials to be estimated reliably.

8.4.1 INDUCTIVE EFFECTS

Three sources of variation in the ionization potentials of lone-pair electrons can be distinguished:
  (1) conjugative (bonded or non-bonded) interaction with other orbitals in the molecule, which may either raise or lower the ionization potential;
  (2) spin–orbit coupling in halogen lone-pair ionizations, which should split the ionization band symmetrically;
  (3) local electric charges and the electrostatic effect of near-by dipoles; these should shift the ionization band bodily, and they constitute the inductive effect which we want to examine.
The spin–orbit splitting in halogen lone-pair ionizations can be allowed for by averaging the ionization potentials for the two peaks, but conjugative and inductive effects cannot be separated in such a simple manner. The strength of conjugative interactions can be estimated to some extent from the breadth of the lone-pair ionization bands, as conjugation gives bonding properties to the otherwise non-bonding electrons. This test is of little practical use, however, because many lone pairs are angle-determining, and

ionization from them gives broad spectral bands for that reason. In large molecules with non-rigid skeletons, the simple presence of a charge may cause changes in molecular geometry by polarization effects, and again give broad bands. Despite the impossibility of separating different effects quantitatively, correlations are found between the ionization potentials of heteroatom lone pairs in homologous series, between ionization potentials and substituent constants and between ionization potentials and electronegativities. The test that is most likely to reveal a correlation is to plot the ionization potentials of lone pairs in one homologous series against those of different lone pairs in another, always taking the same alkyl moieties together. One hopes that the interactions of the alkyl groups with the heteroatoms will be the same for both series, and that the correlation will therefore afford a consistent measure of the substituent effects. Correlations of this nature have been made between many pairs of homologous series using ionization potentials

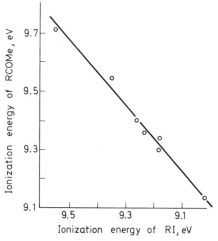

Figure 8.8. Correlation between the lone-pair ionization energies of the methyl ketones and those of the alkyl iodides. (From Cocksey, Eland and Danby[42], by courtesy of the Chemical Society)

from photoelectron spectroscopy[33, 42]. *Figure 8.8* shows, as an example, a graph of the oxygen lone-pair ionization potentials of the methyl ketones against the iodine lone-pair ionization potentials of the corresponding alkyl iodides, and a close correlation can be seen.

Because the ionization bands of the alkyl iodides are sharp, it is convenient to use them as a reference and to plot the ionization potentials in other series against them. The slopes of such plots are a measure of the sensitivity of the heteroatom ionization potentials to substituent effects, relative to the sensitivity of the iodine lone pairs; some values of the slopes are given in *Table 8.1*. The sensitivities increase on going from iodides to bromides and from thiols to alcohols and are greater for lone pairs on atoms adjacent to the alkyl groups than for those which are farther away. The π electron ionizations in the cyanides are the most sensitive of all, while the nitrogen lone pairs in the same compounds are only moderately sensitive.

Table 8.1   SENSITIVITIES OF ADIABATIC IONIZATION POTENTIALS TO ALKYL SUBSTITUTION

Separate figures are given for the nitrogen lone-pair ionizations (RCN-n) and the π orbital ionizations (RCN-π) of the alkyl cyanides.

Series RX	Sensitivity, χ_{RX}	Ionization potential of methyl homologue, I_{MeX}, eV
RNCS	~ 0.57	9.25
RNH$_2$	0.66 $\pm$0.08	8.97
RCO·t-Bu	0.84 $\pm$0.03	9.14
RSH	0.94 $\pm$0.08	9.44
RI	1.00 (standard)	9.54
HCO$_2$R	1.00 $\pm$0.02	10.81
RNO$_2$	1.00 $\pm$0.03	11.08
RCOMe	1.12 $\pm$0.05	9.71
RCN-n	~ 1.2	13.14
RBr	1.25 $\pm$0.03	10.53
RCHO	1.40 $\pm$0.12	10.22
ROH	1.72 $\pm$0.28	10.85
RCN-π	2.00 $\pm$0.02	12.22

In order to ascertain whether the substituent effects are mainly inductive or not, it is necessary to relate shifts in ionization potential not only to one another but also to established substituent constants obtained from physical organic chemistry. The parameters most commonly used for this purpose are Taft's[43] polar substituent constants, σ^*, or Taft and Lewis's[44] inductive substituent constants, σ_I, and several relationships between ionization potentials and these parameters have been proposed that are based on electron impact or photoionization values of ionization potentials[45, 46]. The advantages of ionization potentials measured by photoelectron spectroscopy in the study of substituent effects are their greater accuracy, the fact that either vertical or adiabatic values can be used consistently and that the variations of inner ionization potentials as well as first ionization potentials can be examined.

The relationship between the lone-pair ionization potentials of the aliphatic alcohols and Taft's σ^* constants was examined by Baker et al.[33], who concluded from the scatter of the points that the variations in ionization potential are not completely represented by inductive effects. In a different procedure[42], the ionization potentials of molecules belonging to several series were used to derive values of a new substituent parameter, μ_R. The parameter is defined by the equation

$$I_{RX} = I_{MeX} + \mu_R \chi_{RX} \qquad (8.3)$$

where χ_{RX} are the sensitivity parameters already defined and given in *Table 8.1*. The values of the substituent parameter μ_R for the

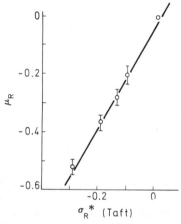

Figure 8.9. Correlation between the substituent parameters, μ_R, defined by equation 8.3 and Taft's substituent constants, σ^*. (After Cocksey, Eland and Danby[42], by courtesy of the Chemical Society)

alkyl groups are given in *Table 8.2*; together with the data of *Table 8.1*, these values allow a large number of ionization potentials to be estimated by using equation 8.3.

The new substituent parameters μ_R specifically represent the effects of different alkyl groups on ionization potentials, and when they are compared with Taft's σ^* constants a close correlation is found[42] (*Figure 8.9*). This correlation shows that although they are not the only factor, inductive effects play a major role in the influence of alkyl substitution on ionization potentials.

Another correlation has been found by Baker et al.[47], who plotted the first and second ionization potentials of the halogen acids and

Table 8.2 ALKYL GROUP SUBSTITUENT PARAMETERS, μ_R

Group	μ_R
Methyl	0.0 (standard)
Ethyl	-0.20 ± 0.03
n-Propyl	-0.29 ± 0.03
n-Butyl	-0.34 ± 0.04
n-Pentyl	~ -0.35
Isopropyl	-0.36 ± 0.02
Isobutyl	-0.38 ± 0.04
t-Butyl	-0.52 ± 0.02

also the lone-pair ionization potentials of the halobenzenes against the Pauling electronegativities of the halogens. They obtained good straight lines and were able to predict from their diagram the σ electron ionization potential of HF, which was uncertain at the time; the predicted value was later experimentally confirmed. This illustrates the main value of the empirical correlations discussed in this section, which is that they allow the ionization potentials of many compounds to be represented by simple empirical formulae and unknown ionization potentials to be reliably estimated.

8.4.2 SUBSTITUENT EFFECTS IN AROMATIC COMPOUNDS

In an important early paper, Baker, May and Turner[48] presented the photoelectron spectra of a large number of monosubstituted and *para*-disubstituted benzenes and interpreted them in terms of substituent effects. Substituents such as $-OH$, $-OR$ and $-NR_2$ with large $+M$ effects, that is, electron-donating by a conjugative mechanism, were found to split the first band in the benzene spectrum into two parts. This band corresponds to ionization from the degenerate orbitals π_2 and π_3, which form the orbital e_{1u} in D_{6h} symmetry. When a substituent is introduced, the degeneracy is lifted, and of the two new π orbitals that result, a_2 has a node at the substituent position and cannot interact with the substituents whereas the other, b_1, can interact with them:

$$a_2 \qquad b_1$$

Of the two new π ionization bands produced in the spectrum, one is accordingly found to remain near its original position, whereas the other is substantially shifted to lower ionization potential by the introduction of a $+M$ group. On the other hand, substituents considered to have a $-M$ effect, or to be electron withdrawing by an inductive mechanism, do not produce detectable splittings but shift the first band to higher ionization potentials. When two substituents are introduced into the benzene ring at *para*-positions, the experimental splittings and shifts are approximately additive. These observations mean that for the study of substituent effects, the photoelectron spectra of benzene derivatives have the advantage that some conjugative and inductive effects can be distinguished. Halogen substituents exert both effects and produce splittings that increase in the order $F < Cl < Br < I$, whereas the shifts that they produce in the mean ionization potential of the first two bands increase in the opposite order, $I < Br < Cl < F$. The shifts, which are interpreted as inductive effects, are in the same order as the substituent effects of the halogens deduced from classical physico-chemical measurements of rate constants and equilibrium constants[49]. The order found for the splittings, apparently due to conjugative effects, is also found in other spectroscopic studies of substitution, and is called the *spectroscopic* order. The fact that it differs from the physico-chemical order might be an indication that conjugative effects are more important in the ions than in the molecules, and too much reliance should not be put on interpretations based on Koopmans' theorem.

A more drastic form of substitution than has yet been considered is to replace all of the hydrogen in a molecule by fluorine. When this is done, it is found that the σ electron ionization potentials of an unsaturated molecule are shifted much further to higher energies than are the π electron ionizations, and a clear separation between π^{-1} and σ^{-1} ionization bands may be produced in the spectrum. *Figure 8.10* illustrates this phenomenon, which has been called the perfluoro effect. The effect has been studied in detail for several pairs of compounds, both aliphatic and aromatic, by Brundle *et al.*[50, 51], who concluded that it might be useful as an aid to the analysis of photoelectron spectra. Its implications for the physics and chemistry of perfluoro compounds have not yet been examined.

8.5 PHOTOELECTRON SPECTRA OF TRANSIENTS

The electronic structure of free radicals and other transient species is of great importance in chemistry, particularly for an understanding of the rates of chemical reactions. The ionization potentials and

218

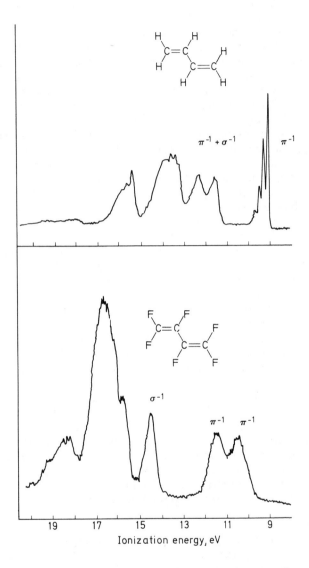

Figure 8.10. Photoelectron spectra of butadiene and perfluorobutadiene. The perfluoro effect is seen in the greater shift of the σ^{-1} than π^{-1} bands to higher
ionization potential on fluorination. The strong band near 17 eV in the spectrum
of the perfluoro compound is due to ionization of the fluorine lone-pair electrons.
(From Brundle and Robin[66], by courtesy of the American Chemical Society)

ionic excited states of radicals are directly relevant to mass spectrometry and to the study of ionic decomposition, for which a knowledge of the energy levels of fragment ions is often needed but seldom available. Photoelectron spectroscopy can provide the necessary information if the transients can be produced in sufficient amounts to give a signal when ionized in the spectrometer source. The usual method is to produce the species of interest by microwave discharge or high-temperature pyrolysis in a flowing gas stream, which transports them as rapidly as possible to the ionization region. Transitory species whose photoelectron spectra have been observed in this manner, mainly by Jonathan and co-workers, include atomic H, N, O, F, Cl and Br[52, 53], the diatomic species SO, CS and P_2[54-57], triatomic NF_2, ClO_2 and S_2O[58-60] and a few larger molecules[61]. The spectra are generally incomplete because either the parent compounds from which the transients are produced, or other fragmentation products from them, are always present in the gas stream together with the species of interest. This causes a masking effect due to overlapping in the spectrum, which is more serious the larger the original molecule. Most of the species mentioned above are rather long lived in a chemical sense and therefore relatively easy to generate and study. Recently, the first ionization band of the more difficult methyl radical has been observed[57, 62], which may be a foretaste of advances to come.

Apart from the photoelectron spectra of transient species in their ground states, those of stable molecules in excited states can be of great interest. The photoelectron spectrum of oxygen in its first electronically excited state, $^1\Delta_g$, has been observed[63], and also some photoelectron peaks ascribed to vibrationally excited nitrogen molecules[52]. In ionization from an electronically excited molecule or atom, it is possible to reach ionic states to which ionizing transitions are normally forbidden by the one-electron excitation rule. This is one approach towards a more complete knowledge of the manifold of electronic states in which molecular ions can exist. It is worth emphasizing that photoelectron spectroscopy usually shows only those ionic states which can be produced by one-electron transitions from the molecular ground state, but these are by no means all of the states in which the ions can exist, even in the energy range below 20 eV.

8.6   ANALYTICAL APPLICATIONS

Photoelectron spectroscopy has a number of characteristics that make it suitable as a tool for qualitative and quantitative analysis, but also some disadvantages in comparison with established physical methods. The balance between favourable and unfavourable factors is at present such that it would be the method of choice only for a limited number of analytical problems.

8.6.1   QUALITATIVE AND QUANTITATIVE ANALYSIS

Photoelectron spectra of different molecules have a degree of complexity and individuality that makes their use as 'fingerprint' spectra possible. With the exception of the spectra of some small molecules, however, they do not consist mainly of clearly separated lines as nuclear magnetic resonance or mass spectra do, but of broad bands that cover wide energy ranges. Even the identification of pure compounds from their photoelectron spectra alone would require the use of intensity data as well as energy data, and this requirement presents a serious problem because of the different energy discriminations in different spectrometers. The identification of compounds in mixtures is even more difficult, as the appearance in the literature of some spectra of mixtures interpreted as those of pure compounds bears witness. Nevertheless, photoelectron spectroscopy may prove to be a useful tool for the analysis of limited groups of compounds, particularly the atmospheric gases and other compounds whose spectra contain easily recognizable features. Compounds that contain chlorine, bromine or iodine have been studied extensively by Betteridge et al.[64, 65] from this point of view, because sharp lone-pair ionization peaks are found in their spectra within narrow energy ranges characteristic of each halogen. The presence of a particular halogen atom in a molecule is indicated fairly reliably by the photoelectron spectrum, with some information about its molecular environment.

For the quantitative analysis of gas mixtures, photoelectron spectroscopy possesses the minimum requirement that band intensities are proportional to the partial pressures of the components. The gases for which analysis by photoelectron spectroscopy is best suited are those whose spectra contain at least one sharp and characteristic peak and do not contain strong extensive band systems, which could obscure sharp peaks characteristic of other compounds. Important among them are N_2, O_2, CO, CO_2, N_2O, CS_2, COS, NO, H_2S, C_2H_2 and all atomic species. A photoelectron spectrum of air showing the resolution of some of these species is shown in *Figure 8.11*.

A real example of quantitative analysis has been given by Daintith et al.[31], who analysed a sample of COS for CS_2 and CO_2 impurities.

Concentrations of about 0.2% of CS_2 in COS would have been detectable in their experiments and this is about the lower limit of detection by photoelectron spectroscopy at present. In the analysis of mixtures of certain gases, photoelectron spectroscopy has an advantage over low resolution mass spectrometry because the photoelectron spectra of the components are different, whereas the mass spectra are almost the same. The pairs N_2–CO, N_2O–CO_2 and CO–CO_2 are the clearest examples of this. Analysis by photoelectron spectroscopy may also be advantageous for compounds whose mass spectra contain no molecular ion peak but only fragment ions. Comparisons with mass spectrometry are especially

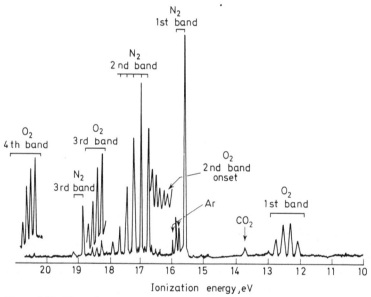

Figure 8.11. Photoelectron spectrum of air. (From Betteridge and Baker[64], by courtesy of the American Chemical Society)

relevant, as it is the technique most closely related to photoelectron spectroscopy in its practical requirements of high vacuum and gaseous samples. In sensitivity and dynamic range, mass spectrometry outstrips photoelectron spectroscopy by several orders of magnitude, needing only microgram instead of milligram samples and offering a dynamic range of $10^5:1$ compared with $10^3:1$. In certain fields, such as upper atmosphere and space research, however, the gases to be analysed are just those for which photoelectron spectroscopy is suited, and the bulk and weight of sector mass spectrometers may be a serious adverse factor. Although it must have strong competition from the lighter forms of mass spectrometer, such as those that work on the radiofrequency, time-of-flight or quadrupole principles, photoelectron spectroscopy would seem to have promising analytical applications in these fields.

8.6.2 STRUCTURE DETERMINATION

If the determination of molecular structure includes electronic structure, then photoelectron spectroscopy is an outstanding technique, but in the determination of ground-state molecular geometries it has many much stronger competitors. A few points of molecular geometry can be determined mainly by photoelectron spectroscopy, but in all instances the aspect of molecular geometry in question must have a strong influence on the molecular electronic structure. The best example is the prediction by Brundle and Robin[66] of the dihedral angle in perfluorobutadiene. Butadiene itself and 1,1,4,4-tetrafluorobutadiene are known to be planar molecules, and the spectra of both contain a gap of above 2 eV between the first and second ionization bands. As the first band in each spectrum definitely represents ionization from the outermost π orbital, the difference in ionization energy between the two π orbitals of each compound must be equal to or greater than 2 eV. In the photoelectron spectrum of perfluorobutadiene, on the other hand, the first two bands, which represent the two π^{-1} ionizations, are separated by only 1 eV (see *Figure 8.10*). This observation suggested that the molecule might not be planar, as in a bent molecule conjugation across the central bond would be less, reducing the splitting between the energies of the two π orbitals. An examination of the optical absorption spectrum in conjunction with the photoelectron spectrum led to the conclusion that the perfluorobutadiene molecule has a *cis*-non-planar configuration with a dihedral angle of 42 degrees, and this conclusion was later fully confirmed by an electron diffraction study. Aspects of the molecular geometry of the biphenyls and the substituted anilines[31] have also been determined by photoelectron spectroscopy.

A striking change in electronic structure occurs in going from a linear to a bent molecule, because orbitals that are degenerate in the linear form become non-degenerate in the bent molecule, as the Walsh diagrams show[67]. The number of bands in the photoelectron spectrum can increase, and the form of the bands also changes very markedly because of the excitation of bending vibrations in ionization from the orbitals which are angle-determining in a bent molecule. If there are any molecules remaining whose linearity is in doubt, photoelectron spectroscopy might therefore be used to establish their structure.

8.7   PHOTOELECTRON SPECTRA OF SOLIDS

Photoelectron spectroscopy has been in use for a long time in the investigation of the electronic structure of solids, particularly metals. A new impetus to this work has been given by molecular photoelectron spectroscopy through the development of intense monochromatic light sources (the resonance lamps) and of improved electron energy analysers.

Because low-energy electrons have a very short mean free path in solids ($< 5\,\text{Å}$), it is essentially the surface of a sample that is examined in ultraviolet photoelectron spectroscopy. It is therefore absolutely essential to have atomically clean surfaces, which means that ultra-high vacuum conditions ($< 10^{-9}$ torr) and very good differential pumping of windowless discharge lamps are required. If the nature of the adsorbed surface layers can be controlled, ultraviolet photoelectron spectroscopy might be a useful method of examining their electronic structure and bonding.

8.7.1   PURE SOLIDS

The photoelectron energy distribution produced by photoionization of a pure solid is a reflection of the electronic band structure. Ideally, the photoelectron spectrum might give the density of states in the different occupied electron bands directly, as the density of states at a particular energy in a full band corresponds to the orbital occupancy in molecular photoelectron spectroscopy. In fact, however, in ultraviolet photoelectron spectroscopy of solids the number of photoelectrons emitted at a particular ionization energy is also a strong function of the final state of the system, that is, of the outgoing electron energy. Photoelectron spectra must therefore usually be taken at several different photon energies before deductions about the band structure can be made. An example of a set of photoelectron spectra is shown in *Figure 8.12*, and a calculated density-of-states function is given for comparison in *Figure 8.13*. A careful comparison shows that peaks in the density-of-states function do correspond to peaks or shoulders in the photoelectron spectra at the same energies relative to the Fermi level, but some of the experimental curves also contain additional peaks, and all have intensity

Ag

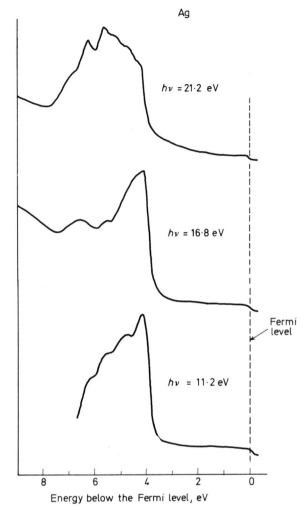

Figure 8.12. Photoelectron spectra of solid silver taken with light of three different wavelengths. The strong bands below 4eV are d electron ionizations. (From Eastman[70], by courtesy of the North Holland Publishing Company)

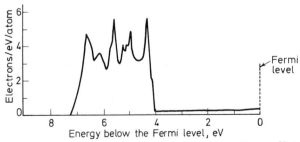

Figure 8.13. Theoretical density of states for silver. (From Eastman[70], by courtesy of the North Holland Publishing Company)

distributions that are different from the calculated curve. The deviation is least at the highest photon energy, and it seems that this trend continues to higher energies. In X-ray photoelectron spectroscopy of metals, the density-of-states function is reproduced directly in the photoelectron energy distribution[68] but at the high electron energies involved it is difficult to achieve the high resolution necessary in order to exploit this simplification.

8.7.2   ADSORBED SPECIES

A great potential exists for photoelectron spectroscopy in the study of species adsorbed on surfaces, about which relatively little is known despite enormous experimental efforts. The photoelectron spectrum of a surface with adsorbed layers compared with the spectra of the free gaseous adsorbate and the clean surface may show up the changes in electronic structure both of the solid and of the adsorbed molecules. This information is being sought by several other techniques as it is clearly crucial for an understanding of the bonding between absorbate and surface, and limited success has been achieved[69]. A start on the study of adsorbed layers by ultraviolet photoelectron spectroscopy has already been made[70, 71], and a large increase in the efforts expended on it is to be expected in the near future. A strong inducement for advance comes from the hope of gaining a better understanding of chemical reactions at surfaces, especially catalytically induced reactions, which make up a large part of industrial chemistry.

REFERENCES

1. LLOYD, D. R. and SCHLAG, E. W., *Inorg. Chem.*, **8**, 2544 (1969)
2. EVANS, S., GREEN, J. C., ORCHARD, A. F., SAITO, T. and TURNER, D. W., *Chem. Phys. Lett.*, **4**, 361 (1969)
3. EVANS, S., GREEN, J. C., GREEN, M. L. H., ORCHARD, A. F. and TURNER, D. W., *Discuss. Faraday Soc.*, **47**, 112 (1969)
4. EVANS, S., GREEN, J. C. and JACKSON, S. E., *J. chem. Soc., Faraday Trans. II*, **68**, 249 (1972)
5. HILLIER, I. H., SAUNDERS, V. R., WARE, M. J., BASSETT, P. J., LLOYD, D. R. and LYNAUGH, N., *Chem. Commun.*, 1316 (1970)
6. GREEN, J. C., KING, D. I. and ELAND, J. H. D., *Chem. Commun.*, 1121 (1970)
7. LLOYD, D. R., *Chem. Commun.*, 868 (1970)
8. EVANS, S., HAMNETT, A. and ORCHARD, A. F., *Chem. Commun.*, 1282 (1970)
9. LLOYD, D. R. and LYNAUGH, N., in Shirley, D. A. (Editor) *Electron Spectroscopy*, North Holland, Amsterdam, 445 (1972)
10. ELAND, J. H. D. and DANBY, C. J., *Int. J. Mass Spectrom. Ion Phys.*, **1**, 111 (1968)
11. HILLIER, I. H. and SAUNDERS, V. R., *Molec. Phys.*, **22**, 193 (1971)

12. CRAIG, D. P. and PADDOCK, N. L., *Nature, Lond.*, **181**, 1052 (1958)
13. BRANTON, G. R., BRION, C. E., FROST, D. C., MITCHELL, K. A. R. and PADDOCK, N. L., *J. chem. Soc., A.*, 151 (1970)
14. FROST, D. C., HERRING, F. G., KATRIB, A., MCLEAN, R. A. N., DRAKE, J. E. and WESTWOOD, N. P. C., *Can. J. Chem.*, **49**, 4033 (1971)
15. DIXON, R. N., MURRELL, J. N. and NARAYAN, B., *Molec. Phys.*, **20**, 611 (1971)
16. CRADDOCK, S. and WHITEFORD, R. A., *Trans. Faraday Soc.*, **67**, 3425 (1971)
17. GREEN, J. C., GREEN, M. L. H., JOACHIM, P. J., ORCHARD, A. F. and TURNER, D. W., *Phil. Trans. R. Soc., Lond.*, **A268**, 111 (1970)
18. BAKER, A. D., BETTERIDGE, D., KEMP, N. R. and KIRBY, R. E., *Anal. Chem.*, **42**, 1064 (1970)
19. FROST, D. C., HERRING, F. G., MITCHELL, K. A. R. and STENHOUSE, I. A., *J. Amer. chem. Soc.*, **93**, 1596 (1971)
20. CRADOCK, S. and RANKIN, D. W. H., *J. chem. Soc., Faraday Trans. II*, **68**, 941 (1972)
21. HOFFMANN, R., *Accounts chem. Res.*, **4**, 1 (1971)
22. HOFFMANN, R., IMAMURA, A. and HEHRE, J. W., *J. Amer. chem. Soc.*, **90**, 1499 (1968)
23. BISCHOF, P., HASHMALL, J. A., HEILBRONNER, E. and HORNUNG, V., *Tetrahedron Lett.*, 4025 (1969)
24. HEILBRONNER, E. and MUSZKAT, K. A., *J. Amer. chem. Soc.*, **92**, 3818 (1970)
25. HASELBACH, E., HASHMALL, J. A., HEILBRONNER, E. and HORNUNG, V., *Angew. Chem.*, **81**, 897 (1969)
26. HASELBACH, E., HEILBRONNER, E., MANNSCHRECK, A. and SEITZ, W., *Angew. Chem.*, **82**, 879 (1970)
27. HASELBACH, E. and HEILBRONNER, E., *Helv. Chim. Acta*, **53**, 684 (1970)
28. GLEITER, R., HEILBRONNER, E. and HORNUNG, V., *Helv. Chim. Acta*, **55**, 255 (1972)
29. LINDHOLM, E., *et al.*, *Int. J. Mass Spectrom. Ion Phys.*, **8**, 85, 101, 215 and 229 (1972)
30. COWAN, D. O., GLEITER, R., HASHMALL, J. A., HEILBRONNER, E. and HORNUNG, V., *Angew. Chem.*, **83**, 405 (1971)
31. DAINTITH, J., DINSDALE, R., MAIER, J. P., SWEIGART, D. A. and TURNER, D. W., in *Molecular Spectroscopy*, Institute of Petroleum, London, 16 (1971)
32. LAKE, R. F. and THOMPSON, H. W., *Proc. R. Soc., Lond.*, **315A**, 323 (1970)
33. BAKER, A. D., BETTERIDGE, D., KEMP, N. R. and KIRBY, R. E., *Anal. Chem.*, **43**, 375 (1971)
34. DEWAR, M. J. S. and WORLEY, S. D., *J. chem. Phys.*, **50**, 654 (1969)
35. BISCHOF, P., HASHMALL, J. A., HEILBRONNER, E. and HORNUNG, V., *Helv. Chim. Acta*, **52**, 1745 (1969)
36. HASELBACH, E., HEILBRONNER, E. and SCHRODER, G., *Helv. Chim. Acta*, **54**, 153 (1971)
37. BISCHOF, P., GLEITER, R. and HEILBRONNER, E., *Helv. Chim. Acta*, **53**, 1425 (1970)
38. HEILBRONNER, E., *Israel J. Chem.*, **10**, 143 (1972)
39. HEILBRONNER, E. and MARTIN, H. D., *Helv. Chim. Acta*, **55**, 1490 (1972)
40. BISCHOF, P., HASHMALL, J. A., HEILBRONNER, E. and HORNUNG, V., *Tetrahedron Lett.*, **13**, 1033 (1970)
41. WORLEY, S. D., *Chem. Rev.*, **71**, 295 (1971)
42. COCKSEY, B. J., ELAND, J. H. D. and DANBY, C. J., *J. chem. Soc., B*, 790 (1971)
43. TAFT, R. W., in Newman, M. S. (Editor) *Steric Effects in Organic Chemistry*, John Wiley, New York, 619 (1956)
44. TAFT, R. W. and LEWIS, I. C., *Tetrahedron*, **5**, 210 (1959)
45. LEVITT, L. S. and LEVITT, B. W., *J. org. Chem.*, **37**, 332 (1972)

46. CHARLTON, M. and CHARLTON, B. I., *J. org. Chem.*, **34**, 1882 (1969)
47. BAKER, A. D., BETTERIDGE, D., KEMP, N. R. and KIRBY, R. E., *Int. J. Mass Spectrom. Ion Phys.*, **4**, 90 (1970)
48. BAKER, A. D., MAY, D. P. and TURNER, D. W., *J. chem. Soc., B*, 22 (1968)
49. MURRELL, J. N., KETTLE, S. F. A. and TEDDER, J. M., *Valence Theory*, John Wiley, London, 309 (1965)
50. BRUNDLE, C. R., ROBIN, M. B., KUEBLER, N. A. and BASCH, H., *J. Amer. chem. Soc.*, **94**, 1451 (1972)
51. BRUNDLE, C. R., ROBIN, M. B., KUEBLER, N. A. and BASCH, H., *J. Amer. chem. Soc.*, **94**, 1466 (1972)
52. JONATHAN, N., MORRIS, A., SMITH, D. J. and ROSS, K. J., *Chem. Phys. Lett.*, **7**, 497 (1970)
53. JONATHAN, N., MORRIS, A., OKUDA, M. and SMITH, D. J., in Shirley, D. A. (Editor) *Electron Spectroscopy*, North Holland, Amsterdam, 345 (1973)
54. JONATHAN, N., SMITH, D. J. and ROSS, K. J., *Chem. Phys. Lett.*, **9**, 217 (1971)
55. FROST, D. C., LEE, J. D. and MCDOWELL, C. A., *Chem. Phys. Lett.*, **17**, 153 (1972)
56. JONATHAN, N., MORRIS, A., OKUDA, M., ROSS, K. J. and SMITH, D. J., *Discuss. Faraday Soc.*, **54**, 48 (1973)
57. POTTS, A. W., GLENN, K. G. and PRICE, W. C., *Discuss. Faraday Soc.*, **54**, 65 (1973)
58. CORNFORD, A. B., FROST, D. C., HERRING, F. G. and MCDOWELL, C. A., *J. chem. Phys.*, **54**, 1872 (1971)
59. CORNFORD, A. B., FROST, D. C., HERRING, F. G. and MCDOWELL, C. A., *Chem. Phys. Lett.*, **10**, 345 (1971)
60. MCDOWELL, C. A., *Discuss. Faraday Soc.*, **54**, 297 (1973)
61. KROTO, H. W. and SUFFOLK, R. J., *Chem. Phys. Lett.*, **17**, 213 (1972)
62. JONATHAN, N., *Discuss Faraday Soc.*, **54**, 64 (1973)
63. JONATHAN, N., MORRIS, A., ROSS, K. J. and SMITH, D. J., *J. chem. Phys.*, **54**, 4954 (1971)
64. BETTERIDGE, D. and BAKER, A. D., *Anal. Chem.*, **42**, 43A (1969)
65. BAKER, A. D., BETTERIDGE, D., KEMP, N. R. and KIRBY, R. E., *Anal. Chem.*, **43**, 375 (1971)
66. BRUNDLE, C. R. and ROBIN, M. B., *J. Amer. chem. Soc.*, **92**, 5550 (1970)
67. WALSH, A. D., *J. chem. Soc.*, 2260 (1953) and following papers
68. HAGSTROM, S. B. M., in Shirley, D. A. (Editor) *Electron Spectroscopy*, North Holland, Amsterdam, 515 (1972)
69. HAGSTROM, H. D. and BECKER, G. E., *Phys. Rev. Lett.*, **22**, 1054 (1969)
70. EASTMAN, D. E., in Shirley, D. A. (Editor) *Electron Spectroscopy*, North Holland, Amsterdam, 487 (1972)
71. BORDASS, W. T. and LINNETT, J. W., *Nature, Lond.*, **222**, 660 (1969)

Appendix I

The Names of Electronic States in Atoms, Molecules and Ions

ATOMS

The name of an atomic state contains three pieces of information, which are written in the form of a *term symbol*:

$$^{2S+1}L_J$$

(1) The superscript on the left is the spin multiplicity, which has the value $2S+1$ where S is the total spin angular momentum in units of $\hbar$. If $S = 0$ the multiplicity is 1, and we speak of a singlet state; if $S = \frac{1}{2}$ the multiplicity is 2, giving a doublet state; and $S = 1$ corresponds to a triplet state. S is the result of combining the individual electron spins, which each have an angular momentum of $\pm\frac{1}{2}$. If no electrons are unpaired the state is a singlet, if one is unpaired it is a doublet, and so on.

(2) The capital letter indicates the total orbital angular momentum, L, the result of combining the individual orbital angular momenta of all the electrons. The value of L is indicated by a capital letter following the convention:

$$L = 0 \quad 1 \quad 2 \quad 3 \quad 4 \quad \ldots$$
$$\ S \quad P \quad D \quad F \quad G \quad \ldots \quad \text{alphabetic, excluding J.}$$

If all shells are full, the individual orbital angular momenta cancel and we have an S state; if one electron is unpaired, L is equal to the orbital angular momentum, l, of that electron, 0 for an s electron, 1 for a p electron, 2 for a d electron, etc. When there are two or more electrons not in full shells, their individual angular momenta can combine in various ways, and several different atomic states arise from the same electron configuration. The configuration p^2 outside closed shells, for example, gives terms ^1D, ^3P and ^1S. The details of how electron configurations relate to atomic states are to be found in textbooks of atomic spectroscopy[1] or valence theory[2].

(3) The right-hand subscript, J, is the value of the total angular momentum that results from combining L and S. J can have the values

$$L+S, L+S-1 \ldots |L-S|$$

In a ^3P state, for instance, $L = S = 1$ and the possible J values are 2, 1 and 0, giving $^3P_2$, $^3P_1$ and $^3P_0$ terms. The energy difference between terms with different J values arises from the interactions between the magnetic moments generated by the electron spin and orbital motion, called spin–orbit coupling.

The term symbols described above are derived on the assumption that the spins of all the electrons in an atom combine to form a resultant S, and the orbital angular momenta of the different electrons combine into a resultant L. This is called the Russell–Saunders or LS coupling scheme, and it is valid strictly for light atoms only. In heavy atoms, and in states where two unpaired electrons are in orbitals of very different size, jj coupling may hold, in which the orbital and spin angular momenta of each electron first couple individually, and the resultant j values of the individual electrons combine to give J for the whole atom. Despite this, the Russell–Saunders terms are used whenever possible, even for heavy atoms.

Apart from the term symbol, the name of an atomic state sometimes contains the symmetry of the wave-function to inversion, indicated by another right-hand subscript, g (gerade) for even states and u (ungerade) for odd states. The g or u character of a state is determined by taking the product of the symmetries of the wave-functions of the individual electrons, using the rules $g \times g = u \times u = g$ and $g \times u = u$. The symmetry of s wave-functions is g, p wave-functions are u, d wave-functions are g again, and so on. The ground state of atomic carbon is $^3P_0$, and as this results from two p electrons it is $^3P_{0g}$. Atomic boron has one p electron and so has a $^2P_{\frac{1}{2}u}$ ground state. Finally, another optional extra in atomic state names is the value of the principal quantum number of the outermost shell, written before the term symbol to distinguish states of the same symmetry but different energies. The two metastable states of the helium atom that are most useful in Penning ionization have the electron configuration 1s2s, so the states are 2^1S and 2^3S.

LINEAR MOLECULES

The electronic states of linear molecules (including diatomic) are named by using term symbols similar to those for atomic states. The multiplicity is shown as a left-hand superscript, the total orbital angular momentum Λ as a capital Greek letter, and the total electronic angular momentum Ω as a right-hand subscript:

$$^{2S+1}\Lambda_\Omega$$

The convention for the Λ values is:

$$\Lambda = 0 \quad 1 \quad 2 \quad 3$$
$$\Sigma \quad \Pi \quad \Delta \quad \Phi$$

In centro-symmetric molecules, the g or u character of the electronic wave-function is indicated by an additional right-hand subscript, as in atoms. Σ states must also be distinguished into Σ^+ and Σ^- states according to the symmetry of the electronic wave-function to reflection in any plane containing the molecular axis. All closed shells give Σ^+ states, and a Σ^- state arises only when $\Lambda = 0$ is attained by cancellation of the orbital angular momenta of individual π or δ electrons. If a molecule in a particular state possesses a single unpaired electron, the term is a doublet, and the Λ value is equal to the orbital angular momentum of the unpaired electron, 0 for a σ electron, 1 for a π electron. The great majority of the ionic states produced by photoionization are of this type, so there is a direct relationship between the symmetry of the orbital ionized and that of the ionic state produced. Ionization of a σ_g electron from a closed-shell molecule gives a $^2\Sigma_g^+$ state, ionization from a σ_u orbital gives a $^2\Sigma_u^+$ state and ionization from a π orbital gives a $^2\Pi$ state. This rule does not apply, however, when molecules that already have unpaired electrons are ionized, because usually the resulting ions will have more than one unpaired electron. There will then be several ionic states corresponding to the different possible ways in which the angular momenta of the unpaired electrons can couple.

The term symbols for the electronic states of linear molecules indicate all the important symmetry properties of the electronic wave-functions, but are not sufficient to identify a state completely, as several states may have the same symmetry but different energies. An identifying letter is often written before the term symbol, sometimes with a tilde mark (˜), to give each state a 'personal' name. The ground state is always given the letter X (or X̃) and the excited states are usually named A, B, C, D, etc. in order of increasing energy. Lower-case letters are used to name states of a different multiplicity from those of the main series. Finally, the Ω value is seldom given for Σ states or singlets, where it is simply equal to S or Λ, respectively, and it is not usually quoted even for $^2\Pi$, $^2\Delta$ or higher states unless special attention is to be drawn to it. As examples of the names of linear ion states we can take those of I_2^+ ions. The occupied molecular orbitals of iodine, starting from the innermost, are $\sigma_g^2\,\pi_u^4\,\pi_g^4$, giving I_2 a $^1\Sigma_g^+$ ground state. Ionization from the outermost orbital gives ground state I_2^+ ions in the $X\,^2\Pi_g$ state, which is actually split strongly by spin–orbit coupling into $X\,^2\Pi_{\frac{3}{2}g}$ and $X\,^2\Pi_{\frac{1}{2}g}$. The next ionic states in order of energy are $A\,^2\Pi_{\frac{3}{2}u}$, $^2\Pi_{\frac{1}{2}u}$ and $B\,^2\Sigma_g^+$.

NON-LINEAR MOLECULES

The names of electronic states of non-linear molecules or ions contain the spin multiplicity indicated as a left-hand superscript in the usual way. The rest of the designation depends on the shape of the molecule, and indicates the symmetry properties of the total electronic wave-function. The orbital angular momentum symbol of atomic and linear states is replaced by a capital letter with various subscripts and superscripts, the Mulliken symbol for the irreducible representation to which the electronic wave-function belongs in the molecular symmetry group. The group theory needed to understand this in detail, and to deduce the appearance of wave-functions or orbitals from the symbols, must be studied in another textbook, such as that by Cotton[3]. The orbitals in a non-linear molecule are named with lower-case Mulliken symbols, so the relationship between the name of an orbital ionized and that of the ionic state produced is the same as for linear molecules. Ionization from an a_1 orbital in a closed-shell molecule produces an ion in a $^2A_1$ state. The names of the states and orbitals of linear molecules are, in fact, a special case of the general rule, as the symbols Σ_g, etc., are just the Mulliken symbols belonging to the symmetry groups for linear species.

The symbols are not arbitrary, and it is useful to recognize some of the main clues contained in them. The following are those that the author finds most useful:

A, B are non-degenerate representations, so A, B states and a, b orbitals are non-degenerate;

E states are doubly degenerate;

T (or F) shows a triply degenerate state;

A, A_1, A$'$ or A_1' in combination with g, if applicable, is the totally symmetric representation.

The individual designations X, A, B, C, etc., showing the energetic order are seldom combined with the names of states of non-linear molecules because of possible confusion with the Mulliken symbols, but they can be used if necessary in order to avoid ambiguity. When a molecule has little symmetry, there may be very many states with the same symmetry of the electronic wave-function, so some means of distinguishing them is needed. For the purpose of identifying ionic states observed in photoelectron spectroscopy, the simplest solution is to use the name of the orbital ionized, a lower-case symbol, instead of the capital-letter state symbol. The orbitals in a molecule, whether it is linear or non-linear, are numbered sequentially within each symmetry type. The first orbital of a_1 symmetry in a C_{2v} molecule is called $1a_1$, the second $2a_1$ and the third $3a_1$, and simultaneously the b_2 symmetry orbitals are $1b_2$, $2b_2$, $3b_2$, etc. Normal bands in the photoelectron spectrum of any closed-shell molecule can therefore be identified unambiguously by the names of the orbitals from which ionization takes place instead of by the symbols for the ionic states produced.

REFERENCES

1. HERZBERG, G., *Atomic Spectra and Atomic Structure*, Dover Publications, New York (1944)
2. MURRELL, J. N., KETTLE, S. F. A. and TEDDER, J. M., *Valence Theory*, John Wiley, London (1965)
3. COTTON, F. A., *Chemical Applications of Group Theory*, John Wiley, New York (1963)

Appendix II

Important Constants and Conversion Factors

CONSTANTS

Symbol	Constant	Value
c	speed of light in vacuo	2.9979×10^{10} cm s^{-1}
h	Planck's constant	6.6256×10^{-27} erg s
$\hbar$	$h/2\pi$	1.0545×10^{-27} erg s
e	electronic charge	4.8030×10^{-10} e.s.u.
		1.6021×10^{-19} C
m_e	electron rest mass	9.1091×10^{-28} g
k	Boltzmann constant	1.3805×10^{-16} erg K^{-1}
N	Avogadro number	6.0225×10^{23}
R	infinite Rydberg	13.6053 eV
amu	atomic mass unit	1.6604×10^{-24} g

CONVERSION OF ENERGY UNITS

1 eV $= 23.061$ kcal mol$^{-1} = 8065.75$ cm^{-1}
1 eV $= 1.6021 \times 10^{-12}$ erg molecule$^{-1} = 9.649 \times 10^{4}$ J mol^{-1}
1 cm$^{-1} = 1.23981 \times 10^{-4}$ eV
1 a.u. (Hartree) $= 27.2107$ eV

UNITS OF WAVELENGTH

1 Å $= 10^{-8}$ cm $= 0.1$ nm
1 nm $= 10$ Å $= 1$ mμ (millimicron)

UNITS OF PRESSURE

1 torr $\quad = 1$ mm Hg $= 1.316 \times 10^{-3}$ atm

1 N m^{-2} $= 10$ dyn cm^{-2} $= 7.5006 \times 10^{-3}$ torr

1 atm $\quad = 1.01325 \times 10^5$ N m^{-2} $= 1000$ bar

Index

Acetylene
  empirical molecular orbital
    calculations, 100–102
  photoelectron spectrum, 12
Adiabatic ionization potentials, 84
Alkyl halides, 142
Ammonia
  fluorescence autoionization, 66
  photoelectron spectrum, 12
Analysis
  chemical, qualitative and quantitive,
    220–222
  chemical structure determination, 222
  chemical, uses of photoelectron
    spectroscopy, 219–223
  photoelectron spectra, of. *See*
    Photoelectron spectra
Angular distribution of photoelectrons,
    66–75
  and electron spin, 73–74
  effect of autoionization, 69–70
  effect on analysis of spectra, 66
  energy dependence, 69
  form of, 67–70
  in vibrational excitation, 72
  in rotational transitions, 74–75
Anharmonicity, 116–118
Anisotropy parameters, 68–72
  table of β values, 72
Anthracene
  molecular orbital calculations,
    102–104
  photoelectron spectrum, 102

Argon as calibrant, 49
Atoms
  photoelectron spectra of, 4–7
  spin–orbit coupling, 6
  term symbols, 228–229
Autoionization, 59–65
  and angular distributions, 69
  and band intensities, 63–65
  fluorescence autoionization, 66

Band intensities, 13–16
  and orbital character, 16–18
  in X-ray photoelectron spectroscopy,
    17
  limiting rules for, 13–16
  measurement of, 50–52
Benzene
  ab initio SCF calculations, 94–99
  photoelectron spectrum, 95
Bicyclo-[2.2.2] octane, photoelectron
    spectrum, 206
Biphenyl, partial photoelectron
    spectrum, 14
Boron triiodide, second-order spin–
    orbit splitting, 138, 139
Butadiene, photoelectron spectrum, 218

Calibration, spectra, of, methods, 49–50
Carbon dioxide
  kinetic energy spectrum, 191–192
  spin–orbit coupling, 136